Helmut Günzler (Ed.)

Accreditation
and Quality Assurance
in Analytical Chemistry

W0055310

Springer

Berlin
Heidelberg
New York
Barcelona
Budapest
Hong Kong
London
Milan
Santa Clara
Singapure
Paris
Tokyo

Helmut Günzler (Ed.)

Accreditation and Quality Assurance in Analytical Chemistry

Translated by Gaida Lapitajs

With 53 Figures and 11 Tables

 Springer

Editor:

Prof. Dr. Helmut Günzler
Bismarckstr. 4
D-69469 Weinheim

Translator:

Dr. Gaida Lapitajs
Schwedenweg 34
D-85560 Ebersberg

Title of the German Edition:
Akkreditierung und Qualitätssicherung
in der Analytischen Chemie.
Original published by:
© Springer-Verlag Berlin Heidelberg New York, 1994

ISBN 978-3-642-50081-7   ISBN 978-3-642-50079-4 (eBook)
DOI 10.1007/978-3-642-50079-4

Die Deutsche Bibliothek – CIP-Einheitsaufnahme

Accreditation and quality assurance in analytical chemistry:
with 11 tables / Helmut Günzler (ed.). Transl. by Gaida
Lapitajs. – Berlin ; Heidelberg ; New York ; Barcelona ;
Budapest ; Hong Kong ; London ; Milan ; Santa Clara ;
Singapure ; Paris : Tokyo : Springer, 1996
Dt. Ausg. u.d.T.: Akkreditierung und Qualitätssicherung in der
Analytischen Chemie
NE: Günzler, Helmut [Hrsg.]

© Springer-Verlag Berlin Heidelberg 1996
Softcover reprint of the hardcover 1st edition 1996

Foreword

Chemical measurements of various kinds are playing a rapidly expanding role in modern society and increasingly form the basis of important decisions. On the basis of these decisions national and international regulations with commercial, legal, medical, and environmental impact are set in force. Therefore, the reliability of chemical measurements is of utmost importance. No acceptance of these regulations without trustworthy decisions and no trustworthy decisions without reliable measurements.

The quality of traded products and therefore their value in trade, depends crucially on the measurements which determine the degree of quality. Furthermore, trust in quality can only exist on the basis of trust in the measurements in proving this quality. On the international scene, acceptance of quality is a transnational problem: how can one trust claimed quality? The Council of the European Union recognized that the differences in the national accreditation systems of laboratories were an obstacle to fair trade and that a harmonization of the national accreditation systems was needed through the introduction of uniform and universally binding criteria.

Without doubt, transnational economical entities such as NAFTA and EFTA will (have to) come to the same conclusion. Thus the EN 45000 series was born from existing systems for national accreditation of test laboratories and many branches of the economy have set up appropriate accreditation agencies.

Chemical compounds and products, however, have also invaded the environment either unwillingly at the endpoint of their industrial life cycle e.g. CFCs, or by accident e.g. oil spills at sea, or as waste e.g. dioxins, radioactive gases, or on purpose e.g. fertilizers and tanker cleaning. Nature can cope with trace amounts of non-natural chemicals, but the extent to which modern society and economy can release chemical compounds into the soil, the water and the air as well as into the human body, sometimes exceeds the absorptive power of the natural systems or, worse, disturbs natural equilibria. In a number of cases, regulation is not only unavoidable but mandatory.

Regulation means setting limits. Verifying the implementation, i.e. compliance with these limits, means chemical measurements. Since chemical pathways in the environment totally ignore man-made borders, chemical compounds in the environment cross these and become, by definition, border-crossing problems. Decisions to cope with these problems must be acceptable across borders, i.e. internationally. Hence, compliance with the necessary international regulations becomes a matter of (trust in) reliable chemical measurements as the basis of these decisions. Quality and reliability in chemical measurement are again of paramount importance.

Never before, have so many discussions taken place around quality in chemical measurement and have so many formal procedures been developed to fulfil the wish to demonstrate quality. Assuming reliability because the chemical measurement was "scientific" seems to be being replaced gradually by assuming lack of reliability unless proven otherwise. The time that chemical measurements were little more than declarations of isolated figures, is definitely over.

Measurements of high quality are not possible without qualified analysts in the field. This book is intended to help the analysts-in-the-field and to support their attempts to be well-informed and to continuously shape and improve the quality of their measurement work. It is not sufficiently recognized that analysts pursue an important societal task by achieving reliable and accepted chemical measurements. It is an explicit purpose of this book to assist the chemical analyst in this task.

Professor Dr. P. De Bièvre
Chairman EURACHEM

List of Contributors

Berdowski, Janusz B.
Polski Centrum Badan i Certyfikacji (PCBC)
ul. Klobucka 23 A, PL-02699 Warszawa, Poland

Berghaus, Hartwig
Director of the Economics Department of the permanent representation
of the Federal Republic of Gemany to the European Union,
Rue J. De Lalaing 19 – 21, B-1040 Brussels, Belgium

Boldyrev, I. V.
Association "Analytica"
9, Leninski Prospect, 117049 Moscow, Russia

Böshagen, Ulrich
EBM Trade Association, Postfach 32 12 30, D-40427 Düsseldorf, Germany

Carlsson, Christina
Centre for Metrology and Accreditation – FINAS; P.O.Box 239,
FI-00181 Helsinki, Finland

Cofino, Wim P.
Free University, Institute for Environmental Studies, De Boelelaan 115,
NL-1081 HV Amsterdam, The Netherlands

Danzer, Klaus
Institute for Inorganic and Analytical Chemistry, Friedrich-Schiller-Universität
Steiger 3, D-07743 Jena, Germany

De Bièvre, Paul
CEC-Joint Research Centre, Institute for Reference Materials
and Measurement (IRMM), Steenweg op Retie, B-2440 Belgium

Del Monte, Maria Grazia
CNMR – Centro Nationale per i Materiali di Riferimento,
c/o Centro Sviluppo Materiali, Via di Castel Romano 100 – 102,
I-00129 Roma, Italy

Dobkowski, Zbigniew
Industrial Chemistry Research Institute, 8, Rydygiera str.,
PL-01-793 Warsaw, Poland

Freedman, G. I.
All-Russia Institute of Light Alloys
9, Leninski Prospect, 117049 Moscow, Russia

Galsworthy, David
UKAS Executive, National Physical Laboratory
Teddington, Middlesex TW11 0LW, UK

Griepink, Bernard
Commission of the European Communities, Rue de la Loi, 200,
B-1049 Brussels, Belgium

Günzler, Helmut
EURACHEM/Germany, Bismarckstraße 4, D-69469 Weinheim, Germany

Ischi, Hanspeter
Swiss Accreditation Service, Lindenweg 50, CH-3084 Wabern, Switzerland

Jensen, Hanne
Novo Nordisk A/S, Building 9G2.10, DK-2880 Bagsvaerd, Danmark

Karpov, Yu. A
Association "Analytica"
9, Leninski Prospect, 117049 Moscow

Koch, Karl Heinz
Technical University Vienna, Institute for Analytical Chemistry,
Getreidemarkt 9/151, A-1060 Vienna, Austria

van de Leemput, Peter
NKO/STERIN/STERLAB – Nederlandse Stichting voor de Erkenning,
PO Box, NL-3001 GD Rotterdam, The Netherlands

Lundgren, Björn
Swedish National Testing and Research Institute,
Box 857, S-50101 Boras, Sweden

Mechelke, Georg J.
Deutsche Akkreditierungsstelle Mineralöl GmbH (DASMIN),
Kapstadtring 2, D-22297 Hamburg, Germany

Mourrain, Jeanne
Groundwater Protection Division, US Environmental Protection Agency,
401, Main Street
USA-SW Washington DC 204602

Quevauviller, Philippe
CEC-Directorate General for Science, Research and Development,
Rue de la Loi, 200, B-1049 Brussels, Belgium

Ramendik, Gregory I.
Institute of General and Inorganic Chemistry, Russian Academy of Sciences
31, Leninski Prospect, 117907 Moscow, Russia

Schlesing, Hendrik
In der Speith 19, D-56235 Ransbach-Baumbach, Germany

Saeed, Khalid
Senior Engineer, Dept. Norwegian Accreditation, National Measurement
Service, P.O.Box 6832 St. Olavs Plass, N-0130 Oslo, Norway

Slyton, Joseph L.
Groundwater Protection Division, Environmental Protection Agency,
401 Main Street, Washington, DC, USA

Trovato, E. Ramona
Groundwater Protection Division, Environmental Protection Agency,
401 Main Street, Washington, DC, USA

Wegscheider, Wolfhard
University of Mining and Metallurgy, General and Analytical Chemistry,
Franz-Josef-Straße 18, A-8700 Leoben, Austria

Contents

1	**Significance of Certification and Accreditation Within the European Market** *Hartwig Berghaus*	1
1.1	Introduction	1
1.2	The EC Commission's Global Concept for Testing and Certification	1
1.2.1	From the New Conception for Harmonization and Standardization Towards the Global Concept	1
1.2.2	The Philosophy of the Global Concept (Building Confidence)	2
1.2.3	Instruments for Building Confidence	3
1.3	The Result of the Council's Discussion on the Global Concept	4
1.3.1	On the Council's Resolution of December 21st 1989	4
1.3.2	The Council's Modular Resolution of December 13th 1990 . . .	5
1.4	The Certification Contents of the New EC Harmonization Guidelines	7
1.4.1	Practising the Modular Concept	7
1.4.2	The Role of the Notified Body	7
1.4.3	CE Labelling	9
1.5	The Influence of the EC Commission's Global Concept on the Private Sector	10
1.5.1	Foundation of EOTC	10
1.5.2	Further European Groups in the Fields of Certification and Accreditation	11
1.6	Assessment of the European Certification/Accreditation Policy	12
2	**The Accreditation of Chemical Laboratories** *U. Böshagen*	15
2.1	The European and International Framework	15
2.1.1	Introduction	15
2.1.2	Free Trade in Europe	15
2.1.3	The Global Concept for Testing and Certification in Europe . .	16
2.1.4	Testing, Certification and Accreditation	17
2.1.5	Objectives of Certification	18
2.1.6	Harmonized Testing Procedures	18
2.1.7	The Significance of the Evaluation of Accreditation Systems	19

2.2	The Basic Principles and the Actual Description of the European and International Framework	19
2.2.1	General Standards for Accreditation and Certification	19
2.2.1.1	General Remarks	19
2.2.1.2	The ISO 9000 Series of Standards	20
2.2.1.3	The ISO/IEC Guide 25	21
2.2.1.4	The EN 45 000 Basic Series of Standards	22
2.2.1.5	The Connection between Both Series of Standards	22
2.2.1.6	Quality Management Systems	22
2.2.2	Specific Recommendations for Accreditation and Certification of Chemical Laboratories	23
2.2.2.1	Good Laboratory Practice	23
2.2.2.2	WELAC/EURACHEM Guidelines "Accreditation for Chemical Laboratories"	24
2.2.2.3	CITAC Document	24
2.2.2.4	Other Documents	26
2.2.3	General Scheme for the Accreditation Procedure	26
2.3	The National Accreditation Systems in Europe	28
2.3.1	Laboratoy Accreditation in Austria – *Wolfhard Wegscheider*	28
2.3.2	The Swiss Accreditation Service – *Hanspeter Ischi*	30
2.3.3	The German Accreditation System – *U. Böshagen*	32
2.3.3.1	The BDI Model	32
2.3.3.2	Cooperation between Regulated and Non-regulated Area	35
2.3.4	The Danish Accreditation System – DANAK – *H. Jensen*	36
2.3.4.1	Structure of DANAK in the Area Concerned with Accreditation of Testing and Calibration Laboratories.	37
2.3.4.2	Accreditation of Chemical Laboratories	38
2.3.5	FINAS – The Finnish Accreditation Service: Laboratoy Accreditation – *Christina Carlsson*	39
2.3.5.1	General	39
2.3.5.2	Operation	40
2.3.5.3	Information	41
2.3.6	The Italian Accreditation System – *M. Gracia Del Monte*	41
2.3.7	The Irish National Accreditation Board	46
2.3.7.1	Background	46
2.3.7.2	Scope of Operations	46
2.3.7.3	Statistics	47
2.3.8	The National Accreditation Body of Norway – *Khalid Saeed*	48
2.3.8.1	General History of the Norwegian Accreditation System	48
2.3.8.2	Organisation	48
2.3.8.3	Tasks for Norwegian Accreditation	49
2.3.8.4	Multilateral Agreements (MLA).	49
2.3.8.5	Application and Assessment Procedure for Laboratory Accreditation	50
2.3.8.6	Assessment Team	50
2.3.8.7	Collaboration with Other Accreditation Bodies.	51

2.3.8.8   Accreditation for the Regulated and Non-Regulated Area . . .   51
2.3.8.9   Sector Committees   51
2.3.8.10   Concluding Remarks   51
2.3.9   The Netherlands Accreditation System – P. van de Leemput   52
2.3.9.1   Introduction.   52
2.3.9.2   Objectives and Accreditations   53
2.3.9.3   International Activities   54
2.3.10   The Polish Accreditation System
           Z. Dobkowski, B. Berdowski   55
2.3.11   Russian System for Analytical Laboratories Accreditation
           Yu. A. Karpov, I. V. Boldyrev, G. I. Ramendik and G. I. Freedman   58
2.3.11.1   Accreditation Criteria.   59
2.3.11.2   Accreditation System Structure   60
2.3.11.3   Accreditation Procedure   60
2.3.12   Accreditation of Laboratories in Sweden – Björn Lundgren   61
2.3.12.1   The Accreditation Body.   61
2.3.12.2   The Accreditation Process   62
2.3.12.3   Types of Laboratory Accreditated.   63
2.3.12.4   Accredditated Laboratories inDifferent Areas (December 1994)   63
2.3.13   The United Kingdom Accreditation System – D. Galsworthy   64
2.3.13.1   History   64
2.3.13.2   The Objectives of NAMAS   64
2.3.13.3   The Formation of UKAS   64
2.3.13.4   Legislative Support for Accreditation in the United Kingdom   66
2.3.13.5   Relationship between Laboratory Accreditation
           and Certification to ISO 9000   66
2.3.13.6   Co-operation between UKAS Laboratory Accreditation
           and the UK Good Laboratory Practice (GLP) Monitoring Unit   67
2.3.13.7   Areas for Development   67
2.3.13.8   Current Concerns   68
2.3.13.9   Case Study: NAMAS Accreditation and the "Additional
           Measures Directive"   68

3   Quality Assurance in Analytical Chemistry
           Karl Heinz Koch   69

3.1   On Quality Assurance   69
3.2   Quality Policy and Quality Management   71
3.2.1   Corporate Quality Policy and Quality Strategy   71
3.2.2   Quality Management and Quality Assurance   72
3.2.3   Total Quality Management (TQM)   73
3.2.4   Quality Costs   74
3.3   Quality Planning, Quality Control, Quality Inspection   75
3.4   Quality Assurance in Analytical Chemistry   77
3.4.1   The Significance of Quality Assurance For and In
           Chemical Analysis   77

3.4.2   Consequences for Quality Assurance in Analytical
          Laboratories   78
3.4.2.1   Compiling a Quality Manual   78
3.4.2.2   Personnel Qualifications and Equipment   81
3.5   QA Measures in Analytical Practice   83
3.5.1   Checking Measuring and Test Equipment   83
3.5.2   Test Control   83
3.5.3   Testing (Test Instructions)   85
3.5.4   Analytical QA Measures   85
3.5.4.1   Control Analyses   85
3.5.4.2   Reference Materials   87
3.5.4.3   Interlaboratory Studies   88
3.5.4.4   Internal Quality Audits   89
3.6   Process Capability and Machine Capability   89
3.7   Certification of Quality Management Systems and
          Accreditation of Analytical Laboratories   91
3.8   References   93

4   **Proper Sampling: A Precondition for Accurate Analyses**
          Wolfhard Wegscheider   95

          Abstract   95
4.1   Sampling Within the Analytical Process   95
4.2   There Is No "Correct" Sampling Without A Clear Problem
          Definition!   96
4.3   Managing Without Sampling?   97
4.4   Planning Sampling Procedures   98
4.5   Aspects of Measurement Uncertainty Caused by Sampling . .   99
4.5.1   Integration Error 100
4.5.2   Materialization Error 101
4.6   Conclusions 102
4.7   References 102

5   **Significance of Statistics in Quality Assurance**
          Klaus Danzer 105

5.1   Types of Errors Associated With Analytical Measurements . . 105
5.2   Systematic Errors 106
5.3   Random Errors 108
5.3.1   Frequency Distributions of Measurement Values 108
5.3.2   Error Propagation 110
5.3.3   Confidence Intervals and Uncertainty Ranges 111
5.4   Significance Tests 113
5.4.1   Tests for Measurement Series 114
5.4.2   Comparison of Two Standard Deviations 116
5.4.3   Comparison of Several Standard Deviations 116
5.4.4   Comparison of Two Means 116

5.4.5   Comparison of Several Means 117
5.5   Statistical Quality Assurance 118
5.5.1   Statistical Quality Criteria 118
5.5.2   Attribute Testing 120
5.5.3   Sequential Analysis 120
5.5.4   Quality Control Charts 122
5.6   Calibration of Analytical Procedures 126
5.6.1   Linear Fit 127
5.6.2   Limit of Decision and Limit of Detection 130
5.6.3   Validation of Calibration Procedures 131
5.7   References 134

6   Validation of Analytical Methods
         Wolfhard Wegscheider 135

         Summary 135
6.1   Introduction 135
6.2   Development of Analytical Procedures and Tasks
         of Basic Validation 136
6.3   Validation: Definitions 138
6.4   Scope and Sequence of Validation 141
6.5   Performance Characteristics 143
6.6   The Relation Between Purpose of the Procedure and
         Scope of Validation 144
6.7   Frequency of Validation 145
6.8   Special Technique of Validation 146
6.8.1   Precision and Trueness 146
6.8.2   Calibration 146
6.8.3   Recovery Studies 148
6.8.4   Comparison of Methods 150
6.8.5   Ruggedness 152
6.9   Conclusions 157
6.10   References 157

7   Traceability of Measurements to SI: How Does It Lead to
         Traceability of Quantitative Chemical Measurements?
         Paul De Bièvre 159

         Preface 159
7.1   Introduction 160
7.2   Traceability of Chemical Measurements: The Problems 162
7.3   Physical and Chemical Measurements: Is There a Difference
         in Principle? 166
7.4   Traceability of Measurements: Are There Precedents? 167
7.5   Traceability of Amount Measurements: Present Status 170
7.6   The "Intersection" Points in a Traceability System 176
7.6.1   Are Reference *Materials* at the "Intersection" Points? 177

7.6.2   How are RMs in Fact Used in Practice?   177
7.6.3   The Real Role of Reference Materials: *Validation*   179
7.6.4   Are Reference Measurements at the Intersection Points? . . .   181
7.6.5   The Place of Reference Materials in a Traceability Scheme . .   181
7.7   Purposes of Traceability of Amount Measurements   184
7.8   Criteria for Traceability of Amount Measurements to the Mole   187
7.9   How can Traceability to the Mole Be Established?   190
7.10   Conclusions   191
7.11   References   193

8   **Reference Materials for Quality Assurance**
          Ph. Quevauviller and B. Griepink   195
8.1   Introduction   195
8.2   Definitions   196
8.3   Requirements for the Preparation of RMs and CRMs   196
8.3.1   Selection   196
8.3.2   Preparation   197
8.3.3   Homogeneity   198
8.3.4   Stability   198
8.3.5   How to Obtain Reference Values   199
8.3.6   How to Obtain Certified Values   199
8.4   The Use of RMs and CRMs in Chemical Analysis   200
8.4.1   The Role of Reference Materials   200
8.4.1.1   The Use of RMs in Statistical Control Schemes   201
8.4.1.2   The Use of RMs in Intercomparisons   202
8.4.2   The Role of Certified Reference Materials   204
8.4.2.1   Calibration   204
8.4.2.2   Achieving Accuracy   205
8.4.2.3   Other Uses of CRMs   205
8.4.3   Suppliers   205
8.4.4   CRMs for Environmental Analysis   206
8.4.5   CRMs for Food Analysis   206
8.4.6   CRMs for Clinical Analysis   207
8.4.7   Other CRMs   207
8.5   References   207

9   **Accreditation and Interlaboratory Studies**
          W. P. Cofino   209
9.1   Introduction   209
9.2   Types of Interlaboratory Studies   209
9.3   Laboratory-Performance Studies in Accreditation Practice . .   210
9.3.1   Objectives of Participation in Laboratory-Performance Studies   211
9.3.2   Assessment of Laboratory Performance   212
9.3.3   The Implementation of Laboratory-Performance Studies . . .   215
9.4   Laboratory-Performance Studies and Quality of Testing . . .   215
9.5   References.   217

10   Accreditation Competence: Requirements
         for Accreditation Bodies
         Georg J. Mechelke 219

10.1   Standard Fundamentals 219
10.2   Organisation and Quality Management System 219
10.3   Arrangements for Accreditation 220
10.4   Operation 221
10.5   Sectoral Committees 221
10.6   Assessment 222
10.7   Assessors 224
10.8   Decision on Accreditation 224
10.9   Diligence and Protective Duties 225
10.10   Surveillance 225
10.11   Accreditation and Standardization 226
10.12   National and International Agreements on Mutual Recognition 227

11   The Significance of Accreditation in Comparison with GLP
         Hendrik Schlesing 229

11.1   Introduction 229
11.2   GLP – Good Laboratory Practice 229
11.2.1   Origin 229
11.2.2   Legal Fundamentals 230
11.2.3   GLP Principles 231
11.2.4   GLP Certificate 235
11.2.5   Personnel 235
11.2.6   Time Needed 235
11.3   Accreditation 236
11.4   Comparison of GLP and Accreditation 237
11.4.1   Quality Assurance 239
11.4.2   Study Plan 239
11.5   Summary and Future Trends 240

12   EURACHEM
         Organization for the Promotion of Quality Assurance in
         Analytical Chemistry and the Accreditation of Analytical
         Laboratories in Europe
         Helmut Günzler 241

12.1   Foundation of EURACHEM 241
12.2   Objectives of EURACHEM 242
12.3   Structural Organization of EURACHEM 242
12.4   Tasks 243
12.5   Cooperation with Other Committees 244
12.6   Summary 245
12.7   References 246

13   **The Accreditation of Environmental Laboratories in the
          United States**
          E. Ramona Trovato, Jeanne Mourrain, Joseph L. Slayton   247

13.1   Introduction   247
13.1.1   Monitoring Systems   247
13.1.2   Challenges   248
13.1.3   Concerns for Data Quality   250
13.2   Policy Development   251
13.2.1   Background   251
13.2.2   Initial Perspectives   252
13.2.3   Assessment of the Need for a National Environmental
          Laboratory Accreditation Program   252
13.2.4   Evaluation of Alternatives to National Environmental
          Laboratory Accreditation   253
13.2.5   Elements of a National Environmental Laboratory
          Accreditation Program   255
13.2.6   Scope of the Program   255
13.2.7   CNAEL's Conclusion and Recommendation   255
13.2.8   Next Steps   255
13.3   Program Development   256
13.3.1   Setting Standards   256
13.3.2   Scope of the Program   256
13.3.3   Federal Role and Responsibility   257
13.3.4   Accrediting Authority Review Board   258
13.3.5   State Implementation   258
13.3.6   Reciprocity   259
13.4   Conclusion   260
13.5   References   260

          Appendix   261

          Subject Index   265

Chapter 1

Significance of Certification and Accreditation Within the European Market

Hartwig Berghaus

1.1
Introduction

The European Communit certification policy plays a key role in the development of the European internal market.

On the one hand, certification policy is part of the harmonization policy, i.e. of the EC-approximation of laws by means of EC directives for harmonization. In fact, apart from uniform quality requirements, the guidelines also include uniform certification procedures. The European certification policy is more than just a harmonization because it also includes the voluntary statutorily non-regulated area.

It is the aim of the European certification and accreditation policy to facilitate the mutual recognition of testing and certificates in order to cut down on costly repetitive testing. Therefore, there is a need for confidence-inspiring measures with a degree of transparency as high as possible.

Apart from theoretical fundamentals of European certification policy, the following describes the practical consequences of this policy for EC member states and third countries.

1.2
The EC Commission's Global Concept for Testing and Certification

1.2.1
From the New Conception for Harmonization and Standardization Towards the Global Concept

In 1985, the Council decided in favour of a new harmonization concept, which is included in the Council's resolution for harmonization and standardization of May, 7th 1985[1]. According to this, EC harmonization guidelines should no longer include, as they had up to this point, all the technical details. They should on the contrary, be restricted to the essential technical requirements and the technical details left to European standardization.

---

[1] OJ No C 136, 4.6.1985.

The "new conception" also includes a special chapter on questions about certification (Appendix II B VIII of the mentioned Council's resolution). This chapter is entitled "Compliance Certificates" (which means the requirements of the particular guideline).

When passing the new conception, the Concil was aware that the contents of its resolution on certification were largely insufficient, and, that, in fact, there was a need for further efforts and decisions. Thus, the Council called upon the Commission to complete the "new conception" in stayes for the assessment of conformity and asked the Commission to deal with this dossier as a matter of priority speeding up all relevant actions.

In order to carry out this request, from 1986, the EC Commission worked on an overall concept for testing and certification. This was completed in July 1989 with the communication to the Council, entitled: "A Global Concept for Certification and Testing" and subtitled: "An Instrument for Quality Assurance of Industrial Products.[2]"

During 1989 and 1990, this document was discussed in the committees of the Council. These discussions led to two decisions. These are:

- the Council' s resolution of 21.12.1989 on a total concept for conformity assessment[3]
- the Council' s decision of 13.12.1990 on the modules for different phases of conformity assessment procedures to be used in the technical harmonization guidelines[4].

1.2.2
The Philosophy of the Global Concept (Building Confidence)

The Global Concept for Certification and Testing aims at facilitating the mutual recognition of testing and certificates even outside the harmonized area. First, these are for product lines not or not yet harmonized and therefore subject to the different quality requirements still existing in the member states and secondly, in the purely private sector, i.e. where only civil law applies. Since the concept includes both the statutorily regulated as well as the voluntary area, the Commission called it the "Global Concept".

According to the EC Commission's communication to the Concil, the mutual recognition of testing and certificates requires confidence: confidence in competence and quality, i. e.

- confidence in product quality
- confidence in the quality and competence of the manufacturers
- confidence in the quality of testing and certification bodies as well as those bodies, that grant accreditation to testing and certification bodies.

Confidence should be fostered by transparency of competence and quality.

---

[2]   OJ No C 267, 19.10.1989.
[3]   OJ No C 10, 16.1.1990.
[4]   OJ No L 380, 31.12.1990.

In the harmonized area, i. e. in the area where EC harmonization guidelines already exist, tests carried out in one member state or certificates issued by one member state are automatically recognised by all other member states. This mutual recognition is in essence, an integral part of each EC harmonization guideline.

Up to this point, confidence in the competence of testing and certification bodies had been taken as granted. According to the Commission, this was an unsatisfactory situation. Therefore, the Commission spoke out for a change. Namely: in future although the designation of national certification bodies in the framework of EC harmonization guidelines should still remain a matter for the individual member states, now, within the framework of the new EC harmonization guidelines passed on the basis of the new conception, the member states should assume the political responsibility for the fact, that the designated testing and certification bodies must continuously meet the minimal requirements stated in the respective appendices of the guidelines.

According to the commission, confidence building was even more important for the not or not yet harmonized areas, i. e. where different legal prescriptions still in the individual member states. The European Court decided in its famous decision entitled "Biological Products"[5] that examinations carried out in a product's country of origin according to the legal requirements of the importing country have to be accepted by the certification body of the destination country. No retesting can be demanded in the destination country, if the testing done in the country of origin is equivalent, and, if the test results are submitted to the certification body of the destination country.

Of course, here the question arises: What is equivalent testing? The European Court did not consider this question in greater detail, however, the EC Commission did. The Commission assumes the conformity requirement has been met if testing is done by accredited testing laboratories on the basis of the relevant internationally recognized assessment criteria.

In the private sector, the mutual recognition of testing and certificates can neither be forced by legal regulations nor by the European Court. Here too, the Commission deems the basic conditions to be a requisite for facilitating the mutual recognition of testing and certificates in the private sector.

1.2.3
Instruments for Building Confidence

Concerning the instruments for building up the necessary confidence and its transparency, in its communication, the Commission starts from the following assumptions:

In the first place, the European standards which the manufacturers should apply to their products are part of these instruments. Manufacturing products according to the European standards, is supported by the assumption that the products fulfill the material requirements of the EC guidelines.

---

[5] legas case 272/80, summary of cases and decisions 1981, p. 3277.

Secondly, to these instruments belong quality assurance measures which should be used by the companies. The international quality assurance standards of the ISO 9000 series have been transfered into European Norms of the EN 29 000 series on the basis of a mandate of the EC Commission. Therefore, they are part of the German national standards (DIN EN 29 000).

The Commission recommends that the manufacturers not only establish internal quality assurances systems, but also to orient them on the European Norms of the EN 29 000 series as well as to get their quality assurance systems certified by an independent third-party for the benefit of transparency.

Thirdly these instruments include the accreditation of testing and certification bodies. The Commission recommends member states, not only to use the instrument of certification as it is, but to establish central accreditation systems in their countries after the models of Great Britain and France.

Accreditation should then be carried out according to the European Norm EN 45 000, drawn up by CEN/CENLEC after appropriate preparatory work done by experts of the member states and the EC Commission with reference to analogous international work by ISO/IEC.

The Commission has in mind that such central accreditation networks of the member states enter into agreements on mutual recognition with the consequence that testing done by testing laboratories which are accredited by their respective national network are assumed to be equivalent. Finally, the Commission proposes in the Global Concept to establish an European infrastructure for certification. This is to promote the conclusion of voluntary agreements on mutual recognition of testing and certificates in the private sector.

This is a summary of the essential contents of the EC Commission "Global Concept for Certification and Testing" of 1989.

1.3
The Result of the Council's Discussion on the Global Concept

What are the consequences which have followed the Global Concept up till now. First, one has to consider the two – already mentioned – decisions, passed by the Council of ministers following the Commission's communication.

1.3.1
On the Concil's Resolution of December 21st, 1989

In its resolution of December, 21st 1989, the Council adopted as a political program certain guidelines of the EC Commission's Global Concept, without a pertinent detailed discussion on the communication of the Commission.

The elements for conformity assessment of the Global Concept, politically consecrated by the Resolution of December, 21st 1989, are summarized as follows:

- The Council recommends the promotion of the general use of quality assurance standards of the EN 29 000 series and its certification.

- The Council, furthermore, recommends the application of the series of standards EN 45 000; these are standards which contain requirements for testing, certification and accreditation bodies.
- The Council favours setting up accreditation systems; i.e. central national systems on the model of the British NAMAS, for example.
- The Council endorses the formation of a testing and certification organization, which should have the following essential tasks: to initiate agreements on the mutual recognition of certification and testing in the non-regulated area and to constitute the framework for it.
- The Council asks the Commission to examine the existing infrastructure of testing and Launch development programs for those countries having an infrastructure below the aspired general community level.
- Finally, the Council expressed its willingness to sign community level agreements on mutual recognition of testing and certificates with third countries.

1.3.2
The Council's Modular Resolution of December 13th, 1990

The second decision of the Council on the basis of the Global Concept of the Commission is the Council's Resolution on "the modules to be used in the technical harmonization guidelines" with regard to different phases of the conformity assessment procedure, issued December 13th, 1990.

Certification modules are certification elements which should be applied in EC harmonization guidelines – and only there. These are, for example, the manufacturer's conformity declaration (module A), the EC model testing (module B) or the EC testing (module F) (cf. scheme on page 6).

Also, a combination of these modules may be applied. For example, a construction type inspection may be fundamentally necessary in combination with the testing of a model. Afterwards, for example, the individual product may be alternatively certified:

- either by testing by a third-party (EC testing) according to modul F, or,
- by the manufacturer himself (subject to the provision that he has a third-party certified quality assurance system available according to modul D or E).

Basically, as defined by the General Guidelines of the modular resolution, EC harmonization guidelines should offer the manufacturer a selection of several alternatives for certification. Furthermore, the types of certification shall be proportional to the risks and dangers etc. intrinsic to the product which is the subject of the guideline. The aim is to avoid "overburdening" the manufacturer.

Conformity assessment procedures in the frame of EC-legislation

ENTWURF (design)

A. (internal inspection of production processes)	B. (type-examination)	G. (inspection of an individual item)	H. (total QA)
manufacturer - puts technical documentation at the disposal of national authorities A.a - calls in the disignated body	manufacturer submits to the designated body - technical documentation - model designated body - verifies conformity with regard to basic requirements - performs tests if appropriate - issues type-examination certificate	manufacturer - provides technical documentation	EN 29001 manufacturer - maintains certified QA system for product design designated body - verifies QA system - checks design conformity[1] - issues certificates on design testing[1]

PRODUKTION (production)

A.	C. (type conformity)	D. (QA production)	E. (QA products)	F. (product testing)	G. (inspection of an individual item)	H. (total QA)
manufacturer - declares conformity with basic requirements A.a designated body - verifies certain product features[1] - sampling inspections[1]	manufacturer - declares conformity with certified type - affixes CE label designated body - verifies certain product features[1] - makes random tests	EN 29002 manufacturer - maintains certified QA system for production and testing - declares conformity with certified type - affixes CE label designated body - recognizes QA system - monitors QA system	EN 29003 manufacturer - maintains certified QA system for monitoring and testing - declares conformity with certified type or with basic requirements - affixes CE label designated body - recognizes QA system - monitors QA system	EN 29003 manufacturer - declares conformity with certified type or basic requirements - affixes CE label designated body - verifies conformity - issues conformity certificate	manufacturer - present product - declares conformity - affixes CE label designated body - verifies conformity with basic requirements - issues conformity certificates	manufacturer - maintains certified QA system for production and testing - declares conformity - affixes CE label designated body - monitors QA system

1 Further regulations may be stipulated by individual guidelines. QA = quality assurance; QA-system = quality assurance system.

1.4
The Certification Contents of the New EC Harmonization Guidelines

1.4.1
Practising the Modular Concept

Since the passing of the new conception in 1985, when the reference system to standards was introduced as a new harmonization method, *twelve* guidelines have been passed. In detail these include: toys, simple pressure vessels, machinery including so-called movable machinery, construction products, electromagnetic compatibility, gas consuming devices, personal protection equipment, non-automatic balances, medical implantations, telecommunication apparatus, boilers, and pharmaceuticals.

All these guidelines contain a voluminous section on certification – sometimes with an appendix of several pages. In each there is a description of how conformity to the essential requirements can be verified (these are mostly safety requirements). These sections on certification of the guideline are part of its "essentials" as is the scope and the essential technical requirements. The General Guidelines included in the Council's modular resolution of December 13th, 1990 as well as the individual certification elements can be found to a large extent in the guidelines, although when the modules were discussed, most of the guidelines had alrady been passed or were being discussed in parallel by committees of the Council.

A manufacturer, fulfilling the requirements of an EC harmonization guideline according to the new conception by using a construction which conforms to a standard the advantage of simplified certification processes. For example, producing toys according to the relevant European standards, the manufacturer's simple conformity declaration is sufficient certification. If he does not manufacture according to the standard, a model testing by an independant third-party is required.

1.4.2
The Role of the Notified Body

What are these notified bodies, which play an essential role for certification procedures in the framework of the harmonization guidelines according to the new conception? They may be certification, testing or surveillance bodies. They often fulfill different functions (type-examination, EC testing or recognition and surveillance of quality assurance). The member states have to designate these bodies and then they are used in the Official Journal of the EC.

Formerly, confidence in the work of these bodies was simply assumed, now the notified bodies have to fulfill certain minimum requirements that are in general identically worded in the guidelines (see for example toy guideline appendic III)[6] (printed on page 13).

If the bodies fulfill the requirements of the standard series EN 45000, it is supposed that they in fact fulfill the requirements of the guideline.

A European manufacturer does not have to have his product tested by a testing body designated by the Government of his country. He may have tests carried our by a testing body in another member state, on condition that it is a notified body. A US-manufacturer, for example, is also free to decide in which member state he wants to have his product tested. However, in each case, it must be by a body designated by a member state. The notified bodies of one member state may offer testing and certification services in the other member states, too. This already follows out of the right of the free movements of services.

This situation may lead to lively competition amongst the European testing and certification bodies which have the status of a notified body. In this competition, apart from the reputation of these bodies, delays and costs for testing will probably play a decisive role.

As yet, there is no final decision on the question as to whether the manufacturer's in-house laboratories may become a notified body. One has to say no, at least as a basic principle because manufacturers might lack the necessary impartiality. The wording of the relevant appendices of the guidelines defining the minimum requirements which have to be fulfilled by notified bodies is relatively strict and does not give much room for interpretation.

It has alrady been mentioned that testing carried out in one member state is effective for all other member states. By the harmonization through EC harmonization guidelines, double and replicate testing is avoided. This is an advantage for the EC-manufacturer as well as for the non EC-manufacturer. The GATT principle for national treatment applies.

In this context, the question arises whether a product of a third country may also be tested for its conformity to the European basic technical requirements by a testing body located in the third country, with the result that further testing in EC-Europe is unnecessary.

The answer is basicaly no. In any case, as long as there is no agreement on mutual recognition of testing and certificates. Partners of such an agreement would be the EC Commission and the respective government of the third country if testing in the harmonized area is to be considered. In this case, individual EC member states – as for example the FRG – may not figure as contracting partners, even when, in actual fact, the product is to be imported only by Germany. In the harmonized sector – without doubt – only one common competence exists. The Commission is basically willing to sign such agreements with third countries. The Council's resolution of 21.12.89, alrady mentioned several times, stipulates the following criteria for such agreements:

– The testing and certification bodies of third countries should have and maintain the same competence as required for testing and certification bodies of community countries.

---

[6]   OJ No L 187, 16.7.1988.

- Mutual recognition is restricted to certificates and marks issued by the bodies mentioned in the respective agreement.
- There must be a balanced ratio with the benefits. These benefits are related to the conformity assessment, not to commercial relation as such.

In the meantime, the Council has given a mandate to the EC Commission to start negotiations with a series of countries. Amongst these are the USA and Japan.

Negotiations with the US American side show the most progress.

Until now, they have been concerned with the product areas of the guidelines following the New Conception. Though, up to now, no understanding has been reached.

Regarding other countries, we are still at the very beginning.

1.4.3
CE Labelling

In industry and commerce insufficient clarity still exists with regard to CE labelling – or, as often said, too: the CE mark.

CE labelling is connected with the technical EC harmonization guidelines following the New Conception. The guidelines passed until now by the Council have already been mentioned in sect. 1.4.1.

Products which do not come within the province of one or more of these guidelines are not affected by compulsory CE labelling.

CE labelling means that the labelled product meets the technical requirements of one or more guidelines of the New Conception and that it has passed the necessary certification procedures; no more, but also not less.

CE labelling is not an indication for quality, in any case, not in the sense of total quality as expected by the consumer, CE labelling only indicates the existence of a certain legally prescribed quality.

This may be explained easily taking a vaccuum cleaner as an example:

A CE-label on a vaccum cleaner implies that the cleaner meets all the requirements of the EMV guideline. From 1.1.95 onward, CE labelling will include the supplementary statement, that the cleaner also fulfills the requirements of the guideline on low voltage, because, starting from this date, CE labelling becomes obligatory under this guideline, too.

However, CE labelling does not provide any information on other quality features which are also very important for the consumer such as, for example, the usuability, the mean time before failure, the design, etc.

Therefore, CE labelling is not consumer-oriented, but is directed towards market supervisory authorities. This labelling is to lighten their task of market supervision. CE labelling is best characterized as an administrative label. It is a sort of passport for the labelled product. The product is supposed to fulfill the requirements prescribed by one or more EC guidelines. The labelling facilitates free trade.

CE labelling is neither assigned nor can one go somewhere and obtain the label. Each manufacturer has to find out himself if his product falls within the scope of one or more guidelines according to the New Conception. Only in this case – as already mentioned – does the product have to be compulsorily labelled. In these circumstances, he has to CE-label the product on his own initiative.

With regard to other labels, it should be mentioned that it is forbidden to attach additional labels which may mislead people with regard to the significance and the typeface of the CE labelling.

However, national standard conformity labels (e.g. also the German RAL quality mark) may be used beside CE labelling.

1.5
The Influence of the EC Commission's Global Concept on the Private Sector

In its Global Concept, the EC Commission proposed the establishment of a European infrastructure for certification. In his already mentioned resolution of December, 21st 1989, the Council accepted these deliberations as its own and recommended founding a testing and certification organization on the European level. It should work in a flexible and nonbureaucratic way. Its essential task should be to bring about agreements on mutual recognition and to constitute the focal point for the drafting of such agreements in the statutorily non-regulated area.

1.5.1
Foundation of EOTC

The idea of such an European testing and certification organization (not accreditation organization) operating in the private sector was realized by the EC Commission in cooperation with EFTA and the European standard organizations CEN and CENELEC. On April 25th, 1990, they signed a Memorandum of Understanding on the foundation of a European organization for testing and certification (European Organization for Testing and Certification – EOTC).

After an experimental phase, terminated at the end of 1992, EOTC was established as an independent organization according to Belgian legislation.

The EOTC has developed its own proceedings as well as guidelines for the sector committees and for the recognition and publishing of agreement groups. The aim is – as mentioned already – to reach mutual recognition of testing and certificates in the non-regulated area.

The most important sector committee is ELSECOM. Under this term, CENELEC has grouped all its certification activities in the field of electrotechnical engineering.

In total, 10 agreement groups have been or are on their way to being developed or recognized.

The EOTC is still mainly supported financially by the EC. When this public financial support is stopped or drastically reduced, the true value of the EOTC for the production and consumer industry will be revealed. The majority of industrial branches have not yet noticed the EOTC. It is the authorities, standard organizations, national accreditation as well as testing and certification bodies and their European associations who control the operative characteristics and politics of EOTC, and only to a very small degree industry, in whose interest EOTC was founded. So, one may follow eagerly the further development of EOTC.

1.5.2
Further European Groups in the Fields of Certification and Accreditation

Following the Global Concept and the foundation of the EOTC, a series of European institutions have been constituted in the field of testing, certification and accreditation. They have developed a lively independance with a paper flood, which as yet only a few national experts can fully comprehend.

In detail, there are the following new foundations:

- WELAC   = Western European Laboratory Accreditation Cooperation (founded 1990)
- EAL   = European Co-operation for Accreditation of Laboratories (founded 1994 by a merger of WELAC and WECC)
- EAC   = European Groups for the Accreditation of Certification (founded 1991)
- EUROLAB   = Organization for Testing in Europe (founded 1990)
- EURACHEM = Cooperation for Analytical Chemistry in Europe (founded 1989)
- EQS   = European Committee for Quality Systems Assessment and Certification (founded 1989)

As well as the already mentioned WECC (Western European Calibration Co-operation) – a cooperation of national calibration services – which has existed since 1975 and the ECITC = European Committe for Information Technology Testing and Certification (founded 1988) are part of this context.

For the sake of completeness two other organizations should also be mentioned though they belong to the statutorily regulated (harmonized) area. These are:

- EOTA   = European Organization for Technical Approvals (founded 1990) as well as
- WELMEC = Western European Legal Metrology Cooperation (founded 1989).

National registration bodies responsible for technical approval according to the guideline on construction products cooperate within EOTA;

WELMEC is the cooperation of national approval and executive institutions in the area of *legal* metrology.

1.6
Assessment of the European Certification/Accreditation Policy

The efforts of the EC Commission and the Council towards a European Certification policy are certainly worthwhile and how they have gone about it deserves our recognition.

In the framework of harmonization by EC harmonization guidelines, the "General Guidelines" and certification elements of the Council's modular resolution of 13.12.1990 constitute reliable fundamental principles for the European legislator. The fundamental clause that the manufacturer has several certification alternatives at his disposal has been substantially realized in the guidelines following the new conception which have been passed up till now.

It is proper that confidence in the testing and certification bodies designated by the individual member states is no longer automatically assumed and that these bodies should fulfill essentially identical criteria defined in the guidelines and that the member states must take over political responsibility for their existence.

On the other hand, the clear tendency to suppressing EC guidelines, the manufacturer's conformity declaration as a type of conformity assessment is questionable. Harmonization guidelines recently passed according to the new conception link the manufacturer's conformity declaration to an existing quality assurance system which is recognized and controlled by an independant third body. This is not an accidental development, it is a reflection of industrial policy goals decided in Brussels. The intention is to extend quality assurance systems and their certification, not only be means of the market but also by EC harmonization.

Certainly, the implementation of internal company quality assurance systems is of interest to companies; avoiding mistakes is cheaper than finding and eliminating them.

Is there, however, a need for complete certification as propagated by the Commission and requested more ad more by the market? In the non-regulated area, this development may lead to new de-facto trade barriers if the existence of a third party certified quality management (assurance) system becomes a prerequisite for market entry. Unfortunately, this development can be observed in several European markets. In Great Britain, for example, there is an increasing tendency that German companies can only successfully market their products which are not subject to public legal product requirements only when they have been subject to a third party certified quality assurance system. Here, agreements on mutual recognition of quality assurance certificates may be helpful but they still have to prove their usefulness. There is a lot of criticism of CE labelling but this does not lie within the frame of reference of this contribution. The European organizations such as EAL, EAC, EUROLAB, EURACHEM etc., recently founded in the shadow of the EOTC, have become so numerous that it is difficult to keep Track of them and so must be seen with a critical eye. A powerful bureaucracy has evolved that is too concerned with itself. Especially alarming is the fact that every organization has either formed or is going to form its own working-groups to deal with similar or even identical themes. This leads to a lively Euro-tourism in the field of testing, certification and accreditation; up to now, at any rate producer and user (consumer) who should profit from the mutual recognition of testing and certification have not received any benefit. Even worse, important branches of European industry are not even aware of these activities.

It seems irrefutable that EAC, EURACHEM, EUROLAB, and EAL will have to combine their activities and even have to merge to some extent. This would increase transparency, cut costs, and, in this way, certainly improve their acceptance. The already successful programme of cooperative activities of WELAC and WECC leading to a merger is a promising start.

Appendix III

Conditions to be fulfilled by the notified bodies (article 9 paragraph 1)

Institutions designated by the member states have to fulfill the following minimum requirements:

1. Have the necessary staff together with suitable materials and equipment;
2. Staff should have technical competence and professional integrity;
3. Independance of executives and technical staff from all groups or persons who have, directly or indirectly vested interests in the toy market, with regard to the performance of test procedures and the documentation of reports, the issue of certificates and surveillance activities according to this guideline;
4. Observance of professional secrecy
5. Taking out a public liability insurance as far as liability is not regulated by the government through national legislation.

The fulfillment of the requirements according to point 1 and 2 is inspected regularly by the responsible bodies of the member countries.

This contribution was published in a similar form in „Zertifizierung und Akkreditierung von Produkten und Leistungen der Wirtschaft", W. Hansen, (ed), 1992, Carl Hanser-Verlag, (3-446-17108-8).

Chapter 2

The Accreditation of Chemical Laboratories

U. Böshagen

2.1
The European and International Framework

2.1.1
Introduction

At the beginning of 1993, the European internal market became formally effective according to the political objectives of the Single European Act (SEA) adopted in 1986. Clause 7a of the SEA spells out the areas which characterize the European internal market. One of which is free trade. As will be shown later, the elaboration of a standard has undergone changes and, therefore, the determination of a product requirement has been subject to alterations, too. An essential objective of the European internal market is to realize compatibility of product requirements and to synchronize the basic components of compatibility, i. e. the work of testing laboratories.

2.1.2
Free Trade in Europe

The Single European Act links the realization of the internal market to 4 essential objectives (clause 7a EEC treaty). One of these is free trade, defined, according to the EC-Commission, by an exchange of goods and services that is largely unrestrained, by the recognition of tests carried out in another country, and, by the confidence of the member states, consumers and purchasers in the testing procedures of other countries. Excluding some isolated and compelling exceptions based on national characteristics, the following principles are now valid within the EC:

There is no reason why, a product, legally manufactured and marketed in one member state, may not be sold without hindrance in another member state of the Community. This testimony to free trade is the original and essential meaning of the CE-mark; nothing more and nothing less. As will be shown later, the original evaluation of the CE-mark developed further to give the CE-guideline which was passed. There are different reasons for trade restrictions and for the emergence of technical obstacles. On the one hand, there are different opinions, regarding the means to be used for the protection of public health, safety or the environment. Thus, standards for certain machinery may call for the employees' own responsibility or may be oriented towards the complete protection against any kind of negligence.

One distinguishes three types of non-tariff restraints:

The first group is based on different national industrial standards which affect the import, sale, or use of a product. These standards concerning performance features, design, function, quality, compatibility or other characteristics of a product prescibed by private sector institutions are not legally binding but "recommendations" according for example, to the standing jurisdiction of the German Federal Supreme Court. Nevertheless, they can be a considerable hindrance to free trade.

Another group of restraints comes from various national regulations, similar to standards, but they have become legally binding together with the standard they are based on.

As a rule, such regulations are enacted for reasons of public welfare, for the protection of health, safety or the environment. For example, in many countries of the Community, the composition of specific foodstuffs is prescribed by law. Commercial exploitation of imported goods that do not meet these requirements contravenes law.

Finally, a third group of restraints is formed by testing methods and certification procedures established to guarantee product conformity to relevant national regulations or industrial standards in their currently updated version. There is an impediment to trade whenever the importing country stipulates for a renewed safety testing or certification in addition to the certification issued in the country of origin. This causes additional costs and an additional expenditure of time, which do not exist in a free market.

The other side of the coin is that on the European internal market, prospects increase for Americain or Japanese manufacturers: they are no longer confronted by 12 single markets, but face one single Common Market with its better sales opportunities. In concrete terms, foreign suppliers do not have to worry about fundamental differences and the national pecularities of the member states of the EC internal market. Certainly, language differences will still remain far into the future. The positive effects of market unification, however, neutralize them to a great extent.

2.1.3
The Global Concept for Testing and Certification in Europe

As already mentioned by Berghaus in Sect. 1.2.2, when outlining the essentials and the directions of the Global Concept for Certification and Testing, the aim of free trade assumes, on the one hand, the same pre-conditions and guidelines for the nature of products. On the other hand, only a common certification policy completes this idea. Standardized testing and administration procedures can be regarded as a basis for confidence in the measures undertaken by any other member state. In its resolution of December 21st, 1989 concerning a total concept for conformity assessment, the Council supports the conclusions of the Global Concept for Certification and Testing – An Instrument for Quality Assurance of Industrial Products.

The Global Concept (Communication of the Commission to the Council on 15.06.1989–89/C 267/03) states that the elimination of technical barriers for products, as planned in the White Paper in the framework of a new strategy, must be complemented by the harmonization and mutual recognition of national regulations and standards. The objectives of the Global Concept are described by the Commission as follows: the verification of product conformity to quality-determining technical specifications is in accordance either to compulsory regulations or to market demand. In the first case, for the protection of health or the environment, for safety reasons etc., authorities insist on proof of conformity from the manufacturer before he markets the products. In the second case, purchasers demand this in business transactions and thus, it becomes contractual and a de-facto criterion.

2.1.4
Testing, Certification and Accreditation

The European testing and certification policy aims at a standardization, transparency and comparability of the testing and certification systems within the member states of the European Union. Therefore, national systems have to be reduced to their essential basic elements; these core elements have to be constituted in a comparable and interchangeable way. The intention is, as shown, not only to design the respective testing and certification process to be streamlined, but, also to avoid retesting in any other country, not only by technical adaptation but – beforehand – by extensive measures to inspire confidence in the testing and certification work of another country.

The Commission of the European Union applies its certification policy to improve and facilitate trade. This is to reduce or to avoid duplicate testing, expensive diversions and supplementary bureaucracy. For example, even if a foreign manufacturer delivers goods which meet the technical standards of the importing country and supplementarily guarantees conformity by a certificate issue by a testing authority of his own country, it is not at all certain that he may readily market his products in the supplied country. In these cases, authorities of the importing country usually urge retesting in their own country and for their national conformity certificate. This increases costs and distorts competitiveness. There are multiple examples, especially in the field of mechanical engineering, pointing to the fact, that the final price of a product is considerably increased when this happers.

Considering this extensive and generally oriented policy for the improvement of trading conditions, the separation and differentiation in the private non-regulated and the statutorily regulated sector play an essential role. The borderline dividing both sectors differs from one member state of the European Union to the other and is traditionally supported on different starting points.

Apart from the CEN-Norms for certification which are oriented on international recommendations (ISO 9000 series), the drafted guidelines themselves provide approaches of product testing for certification. Industry accepts the manufacturer's selfdeclaration as a starting point and as a base for conformity certificates. Here, the manufacturer declares on his own responsibility that his product conforms to the relevant technical regulation in its currently valid version.

With regard to accreditation, the Commission proposed standardizing material, personnel and institutional requirements for all testing authorities in Europe. This is to facilitate product conformity certification and to demand equal prerequisites for product verification. The further aim is, to open up without any supplementary testing – according to clauses 30 and 36 of the EC-treaty – the market of a country to products already tested safety for in another country. Another objective of European policy is to reinforce confidence in tests made in other countries.

2.1.5
Objectives of Certification

As stated in the Global Concept of the then EC-Commission, the objective of certification is to enable harmonized standards to be perfected: testing procedures and the work of a testing body also have to be standardized. Only in this way can product conformity according to relevant standards or other technical safety regulations be guaranteed. The harmonization of standards and the rationalization of testing procedures results in a more cost-efficient handling of conformity procedures, with the aim of avoiding duplicate testing. Providing harmonized instruments for conformity assessment is the approval of a common fundamental principle for their application and facilitates the passing of future technical harmonization guidelines concerning the release of industrial products onto the market. This promotes the implementation of the internal market in the field of free trade. Conformity should be attained without making any regulations which may burden the manufacturer unnecessarily; instead, unambiguous and comprehensive procedures have to be achieved.

2.1.6
Harmonized Testing Procedures

The different testing procedures and testing methods that varied from country to country were an annoyance and a barrier to free trade. Examinations approved in one country were not accepted in the neighbouring country. This led to duplicate testing and to considerable additinal expenditure. In order to avoid these additional costs, it was agreed to standardize and harmonize the fundamentals of testing procedures. These harmonized prerequisites for technical safety testing should lead to test results being comparable and that they need not to be repeated after frontier-crossing.

There are a lot of prerequisites for harmonized testing procedures: the institutional, material, and personnel prerequisites for testing have to be brought into line with each other. There should be exchange of experience including and taking into account the know-how that has been acquired at other institutions. Furthermore, harmonized testing procedures are necessary so that customer and consumer gain confidence in the testing of another country which is an essential requirement for avoiding renewed testing.

2.1.7
The Significance of the Evaluation of Accreditation Systems

As already said, inspiring confidence in the work of another country in the field of testing and certification is an essential aim of the Global Concept and of the policy of the Commission of the European Union. To achieve this primary goal, i.e. confidence and transparency in the accreditation system of another country, working-groups capable of evaluating the specialist criteria of a system, of a certification or accreditation body, and to supervise the necessary harmony of the work were set up in the framework of European cooperation. Experts from another country look at the system and the interrelated structures and base their evaluation on the existing series of standards as well as on guidelines published on this subject which are being constantly improved and extended by European study groups. Evaluation of accreditation systems means the verification of its operational capability and the observance of common basic rules by experts of another country.

2.2
The Basic Principles and the Actual Description of the European and International Framework

2.2.1
General Standards for Accreditation and Certification

2.2.1.1
General remarks

Internationally recognized quality assurance systems have to be implemented by each laboratory that wishes to be accredited. The quality assurance systems are described by the international standards ISO 8402 and ISO 9000–9004, established in 1986–1987.

Testing laboratories applying for accreditation must fulfill general requirements according to the following international and/or regional guides and standards ISO/IEC Guide 25 and EN 45001–45003.

For the laboratory itself the most important are: The ISO/IEC Guide 25 that has a broad international significance (used mainly in non-European countries), and the EN 45001 standard that is used mainly in Europe. Both documents are similar in content.

2.2.1.2
The ISO 9000 Series of Standards

Apart from the fundamental standard series 45000ff for the criteria, the existence and the operation of a testing laboratory, the ISO 9000ff series of standards has become of fundamental importance. Both series of standards are fundamentally different and have to be separated from each other. Nevertheless, later on, a point will be mentioned where a link between both series exists.

Below the ISO 9000 series of standard is mentioned in detail:

ISO 9000 General introductory guide to quality management and quality assurance standards.

ISO 9001 Quality systems-Assurance mode for design development, production, installation and servicing capability

ISO 9002 Quality systems-Demonstration of production and installation capability

ISO 9003 Quality systems-Demonstration of final inspection and test capability

ISO 9004 Quality systems-Guide to generic quality system elements.

Since 1987, the international standards of the ISO 9000 series have been incorporated unchanged into various national standard systems (DIN ISO 9000ff) as well as into the European Norms system under the term EN 29000. No international standard has ever gained such widespread application in such a short time as this one.

The reason is the development of industrial policy in Great Britain. At the beginning of the 1980s, special emphasis was placed on the idea of optimizing quality management systems with governmental aid. This production optimization carried out by the manufacturer internally resulted in a certificate issued by state-approved assessors for quality management systems; certificates for QA systems were born. Optimization of quality management systems involves the improvement of production processes, whereas *in* Germany, advances were made with product quality improvement and with the reduc tion of product defects. These strategies are not contradictory in fact they complement each other. European product liability and its adoption in national law, clearly indicate that absence of product defects is still a standard measure of liability.

Quality encompasses many individual aspects; amongst these are: production process optimization, internal testing for technical rationalization, incorporation of people involved in this process, which gives a preliminary stage of a comprehensive Total Quality Management (TOM).

The documentation of the internal company quality management system is an essential element of the ISO 9000 series. ISO 9001 subdivides this process into 20 points:

Cross-reference list EN 29 000

Clause	EN 29001	EN 29002	EN 29003
Management responsibility	1	2	3
Quality management systems	1	1	3
Contract review	1	1	
Quality in specification and design (design control)	1		
Document control	1	1	2
Testing	1	1	2
Test equipment	1	1	2
Test status	1	1	2
Control of nonconforming product	1	1	2
Corrective measures	1	1	
Purchasing	1	1	
Purchaser supplied product	1	1	
Product identification and traceability	1	1	2
Process control	1	1	
After-sales servicing	1		
Handling, storage, packaging, delivery	1	1	2
Quality records	1	1	2
Internal auditing of the quality system	1	2	
Training	1	2	3
Statistical technques	1	1	2

1 = Full requirement.
2 = less stringent than EN 29001.
3 = less stringent than EN 29002.

There is a worldwide discussion of a more detailed form of this series of standards with respect to branches of industries. However, a majority of the technical committees of ISO as well as European and national standard organizations involved pronounced themselves in favour of preserving the generalized form.

2.2.1.3
The ISO/IEC Guide 25

The ISO/IEC Guide 25, 3rd edition 1990 contains general guidelines for calibration and testing laboratories that simultaneously fulfill requirements of ISO standards series 9000 including the model described by the ISO 9002 standard, since the laboratories operate as producers of calibration and test results. For laboratories performing special tests, for example in such areas as chemistry or informatics, the requirements of the Guide 25 have to be modified and adequately interpreted according to specific features of such special tests.

The ISO/IEC Guide 25 establishes the general requirements for the competence of calibrating and testing laboratories.

2.2.1.4
The EN 45 000 Basic Series of Standards

Focusing upon the question, whether an authorized testing body has been working according to relevant rules, constitutes the core of activities according to the series of standards EN 45000:

- EN 45001 General criteria for the operation of testing laboratories.
- EN 45002 General criteria for the assessment of testing laboratories.
- EN 45003 General criteria for laboratory accreditation bodies.
- EN 45011 General criteria for certification bodies operating product certification.
- EN 45012 General criteria for certification bodies operating quality system certification.
- EN 45013 General criteria for certification bodies operating certification of personnel.
- EN 45014 General criteria for supplier's declaration of conformity.

These standards constitute the foundation for future proof of competency related to testing and certification activities. Implementations are already carried out in both, the governmental and private sector, as is illustrated by various efforts for an improved structural organization of accreditation and recognition. These standards are describing in detail the requirements for laboratory work, and, how equal practice throughout Europe may be achieved for testing procedures.

2.2.1.5
The Connetion Between Both Series of Standards

Although, the European Norm EN 45000 series is related to laboratories and the ISO 9000 series of standards is related to the production process of the manufacturer, both series are interconnected in one single point. If a manufacturer intends to optimize his own production system on the basis of the ISO 9000 series of standards, he has to see if it is worth having the laboratories involved in the production process in house. This is often the case, whether for intermediate inspections or for the final testing of his product. In this case, it is assumed that in a manufacturer's certified quality management system, the laboratories on their part meet international standards. This is the only point where both – normally strictly separated – series of standards converge.

2.2.1.6
Quality Management Systems

The birthplace of quality management systems (QMS) and their certification is Great Britain. There, product testing has been partly substituted by an examination of the product manufacturing system. One of the essential aims of an effective quality management system is the optimization of the manufacturing process. Apart from product testing, guidelines offer in part the possibility to certifying the quality management system for the production process.

It is important to mention, that also an optimized quality management system (in the ISO 9000 series of standards this term replaces the former term quality assurance system, QAS) does not alter the conditions of the manufacturer's liability contained in the CE-guideline on product liability issued on 25.07.1985. Even considering optimized systems, liability remains with the manufacturer or with the importing company. Quality management systems and their optimization and certification play an essential role in the case of production networks and just in time deliveries, i.e. when the customer possibly wants to avoid renewed entry testing of the delivered goods.

2.2.2
Specific Recommendations for Accreditation and Certification of Chemical Laboratories

There are no specific standards, such as ISO and/or regional, e.g., European CEN/CENELEC, for accreditation and certification of chemical laboratories. However, chemical testing has been acknowledged as an area where guidance is necessary. Therefore, special documents have been, and are still, being, developed by international organizations that recommend specific rules for chemical analytical laboratories and/or interpret general standardised criteria.

2.2.2.1
Good Laboratory Practice

It should be mentioned that amongst the member countries of OECD the following guideline document is recommended for the setting up and maintenance of a laboratory quality assurance system:

Good Laboratory Practice in testing chemicals. Final report of the OECD Expert Group on GLP. OECD, Paris 1982. Recent edition: OECD Principles of Good Laboratory Practice and Compliance Monitoring, Numbers 1–6, Paris 1992.

The compliance with the OECD-GLP principles is, in some countries, mandatory for laboratories of health and environmental safety testing and for notification or registration of chemicals, particularly in new ones. GLP principles are applied mainly to analytical chemical laboratories concerned with genetic, clinical, pharmacological, toxicological and other biochemical studies, in particular where laboratory animals are used. There is an increasing interest in harmonisation of ISO and GLP guidelines. Recently (9–10 March 1993), the OECD panel on GLP stated that there is no conflict between GLP and ISO/IEC Guide 25. The GLP regulations can be considered as complementary to the general criteria of EN 45001 standard for specialized chemical labratories.

2.2.2.2
WELAC/EURACHEM Guidelines "Accreditation for Chemical Laboratories"

The document has been developed by a joint EURACHEM/WELAC Working Group as the EURACHEM Guidance Document No. GD1 and the WELAC Guidance Document No. WGD 2, Edition 1, April 1993 (recently denoted as the Document of European Accreditation for Laboratories No. EAL-G4). This document provides specific guidance on the accreditation of chemical laboratories for both assessors and laboratories preparing for accreditation. It has been written in the form of an interpretation of EN 45001 and ISO/IEC Guide 25.

The content of the document:

Part	Section	Ref. to EN 45001	Ref. to Guide 25
1	Introduction		
2	Scope of Accreditation	4 & 12	–
3	Staff	5.2	6
4	Environment	5.3.2	7
5	Equipment	5.3.3	6
6	Reagents	–	8.1
7	Methods and Procedures for Calibrations and Tests	5.4	10
8	Calibration and Measurement Traceability	5.3.3	9
9	Reference Materials and Chemical Standards	5.3.3	9.7
10	Use of Computers	5.3.3 & 5.4	10.7
11	Laboratory Audit and Review	5.4.2	5.3–5.5
12	Sampling, Sample Handling and Sample Preparation	5.4.5	10 & 11
13	Quality Control	–	5.6
14	Measurement Uncertainty	5.4.3	13.2
15	Validation	–	–
16	Bibliography		

Appendix A: Example Quality Audit Checklist
         B: Calibration Intervals and Performance Checks
         C: Use of Computers – General Guidance.

2.2.2.3
CITAC Document

The international guide on quality assurance for chemical analytical laboratories is under development by the CITAC working group. The work started in March 1994 at the PITTCON 94 conference in Chicago. The 3rd draft of February 1995, entitled "International guide to quality in analytical chemistry. An aid to accreditation", CG1, contains the following sections:

1. Aims and objectivs.
2. Introduction.

3. Definitions.
4. Accreditation.
5. Scope.
6. The analytical task.
7. Specfication of analytical requirement (ref. to the ISO 9001 standard).
8. Study plan (ref. to the OECD GLP).
9. Non-routine analysis.
10. Staff (ref. to ISO Guide 25, p. 6; ISO 9001, p. 4.18; GLP, sec. II, p. 1).
11. Environmet (ref. to ISO Guide 25, p. 7; ISO 9001, p. 4.11h,J; GLP, sec. II, p. 3,7.2d(iv)).
12. Equipment (ref. to ISO Guide 25, p. 6; ISO 9001, p. 4.11a–b,d–g, i; GLP, sec. II, p. 4.1, 4.1, 7.2b).
13. Reagents (ref. to ISO Guide 25, p. 8.1; ISO 9001, p. 4.6, 4.10.1.1; GLP, sec. II, p. 4.3, 7.2b).
14. Methods/procedures for calibrations and tests (ref. to ISO Guide 25, p. 10; ISO 9001, p. 4.6, 4.7, 4.9.1, 4.10.2, 4.10.3; GLP, sec. II, p. 7, 8).
15. Measurement traceability and calibration (ref. to ISO Guide 25, p. 9; ISO 9001, p. 4.10.2, 4.11b–c, e–f; GLP, sec. II, p. 4.1.2, 5.1.2, 6.2, 7.2.1a).
16. Reference materials and chemical standards (ref. to ISO Guide 25, p. 9.7; ISO 9001, p. 4.11b; GLP, sec. II, p. 6, 7.2a, 9.2.1c).
17. Computers and computer controlled systems (ref. to ISO Guide 25, p. 10.7; GLP, sec. II, p. 8.3.5).
18. Laboratory audit and review (ref. to ISO Guide 25, p. 5.3–5.5; ISO 9001, p. 4.1.2.2, 4.1.2.3, 4.1.3, 4.17; GLP, sec. II, p. 2.2).
19. Sampling, sample handling and preparation (ref. to ISO Guide 25, p. 10,5, 11; ISO 9001, p. 4.8, 4.15.1–4; GLP, sec. II, p. 3.2–3.5, 6, 7.2a, 10.1.1e, 10.2.1a).
20. Quality control (ref. to ISO Guide 25, p. 5.6; ISO 9001, p. 4.10.2).
21. Measurement uncertainty (ref. to ISO Guide 25, p. 13.2; ISO 9001, p. 4.20; GLP, sec. II, p. 9.2.6c).
22. Validation.
23. Miscellaneous.
24. Bibliography/references.

Appendices
A. Quality audit – Areas of particular importance in a chemical laboratory
B. Calibration intervals and performance checks
C. Use of computers – General guidance
D. CITAC working group
E. National and international quality standards/guides.

This CITAC document CG1 is based on the WELAC/EURACHEM guidelines for accreditation of chemical laboratories (see 2.2.2.2) and it is updated taking new materials or developments, views from ouside Europe, and OECD GLP principles into account.

2.2.2.4
Other Documents

It should also be noted that several national organizations involved in the accreditation and certification procedures are developing their own criteria, rules and requirements specifically for chemical analytical laboratories. For example:

- "Accreditation for Chemical Laboratories", National Measurement Accreditation Service (NAMAS), NAMAS Information Sheet 45, Ed. 1, October 1990.*)
- "Accreditation for Microbiological Testing", NAMAS Information Sheet 31, Ed. 2, May 1992.
- "Chemical Testing – Requirements for Registration", National Association of Testing Authorities (NATA), Australia. NATA 1992.
- "Standard Guide for Accountability and Quality Control in the Chemical Analysis Laboratory" ASTM E 882-87.

2.2.3
General Scheme for the Accreditation Procedure

A general scheme for the accreditation procedure is similar in the various countries since the procedure is based on general international standards and guidelines. The accreditation procedure can be initiated by the labratory management if a quality assurance system is implemented and a relevent documentation, in particular the quality manual, exists. Moreover, the result of an internal audit should indicate that the laboratory is well prepared for accreditation. Then, a contact with the accreditation body (AB) and formal request or application for accreditation is needed. According to the DAR, BSI, NAMAS, PCBC or other accreditation bodies, the following stages can be distinguished in the procedure:

1. Application
2. Assessment
3. Accreditation
4. Surveilance

The detailed routes for each stage are shown in Fig. 1 to 4.

---

*) In 1995 NAMAS altered its name to UKAS.

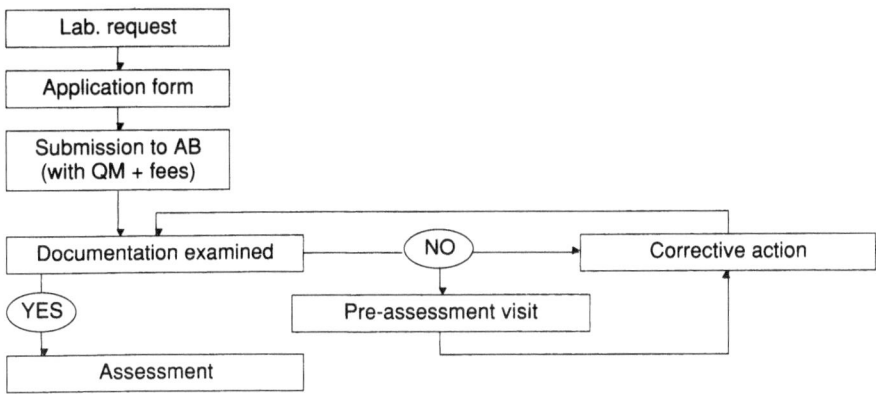

Fig. 1. Application

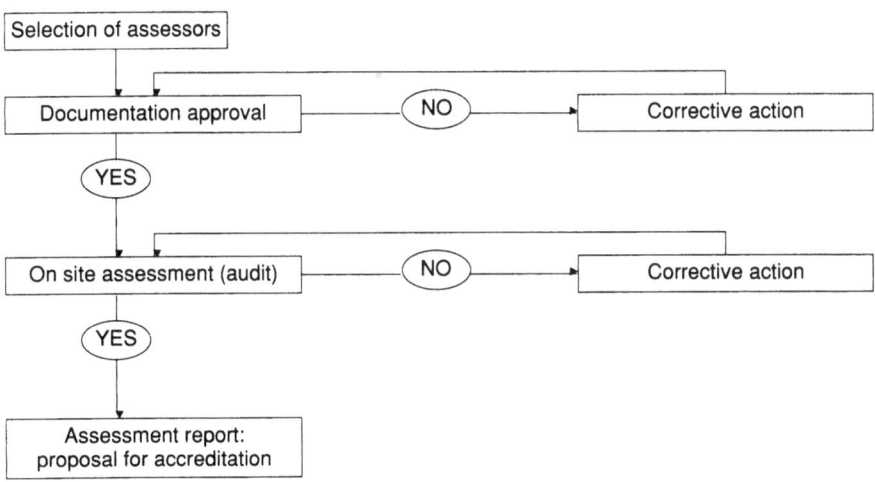

Fig. 2. Assessment

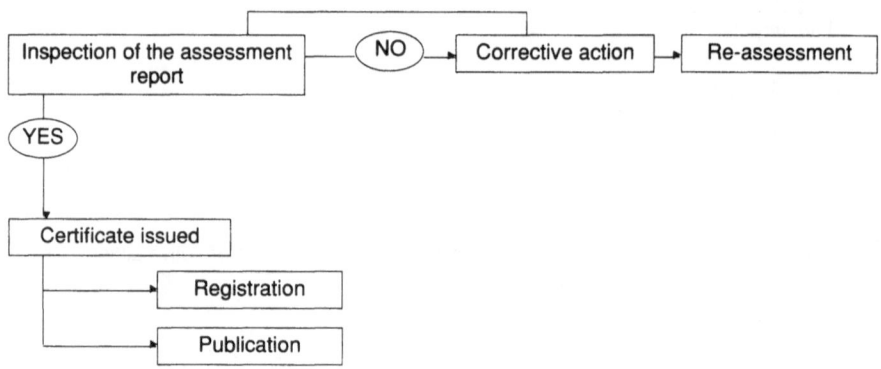

Fig. 3. Accreditation

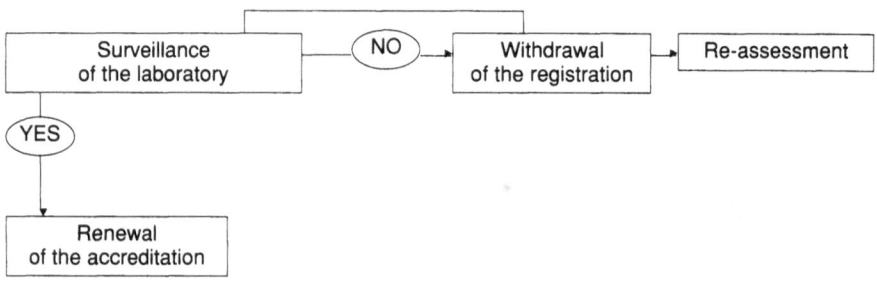

Fig. 4. Surveillance

2.3
National Accreditation Systems in Europe

2.3.1
Laboratory Accreditation in Austria

Wolfhard Wegscheider

The accreditation of chemical (and other testing) laboratories in Austria has several features that can best be understood in terms of the traditional system of "authorized laboratories". This system, which has been in effect since pre-World War I, provided a level of official recognition of test results from selected and inspected laboratories that put test reports issued by these laboratories equal in rank to official documents. As this was a strictly national system, that was, however, highly respected by the public, as well as by the law makers and the courts, the introduction of an EN 45000 Series system in Austria served two purposes:

a) the harmonization on an international level based on ISO Guide 25, and
b) the replacement of the traditional Austrian system of laboratory authorization.

It was mainly for the second reason that accreditation was not introduced in the way other standards are introduced by the Austrian Standards Institute, but by writing it into federal law to make it compatible with the national legal system. There, it has the position of a subsidiary law that takes effect on a voluntary basis and if no other requirements are stated by more specific laws, e.g., in the area of food and pharmacy. The law has been in effect since 1 January 1993. The accreditation body was named in the law as the Austrian Minister of Economic Affairs and the existing authorization body was transformed into one that meets the pertinent EAL/WELAC guidelines. In principle, accreditation in an agricultural context is also done by the Austrian Minister of Agriculture and Forests, but this route is a special and rarely practiced one. In order to meet with the intentions and regulations of the EN 45000-Series, a separate body with substantial independence was established within the Ministry to execute the newly created accreditation law. Additionally a Council of Governors was appointed that advises on important matters on a sectorial basis.

As laboratory authorizations according to the older law have lost their validity, there is also substantial pressure on the laboratories to apply for accreditation. Additionally the system appears also to act as vehicle for public laboratories to gain recognition on a broader basis than just the national one. It is widely felt that irrespective of the current legal situation, accreditation may lead to better international recognition therefore may also be beneficial to public laboratories. By law, the Austrian accreditation system is confined to laboratories within the national boundaries, but they have a higher status in that they are charged with issuing official documents and certificates.

If a laboratory requests accreditation, this request has to include documents such as the quality handbook, a complete list of test procedures for which accreditation is sought and information on the legal status of the laboratory. After formal checks upon receipt, a team of auditors is appointed that is agreed with the Council of Governors. As there is a shortage of auditors in clerical positions the accreditation body often involves auditors from the private sector who are both versed with quality assurance and technical matters.

To ensure maximum compatibility with corresponding systems across Europe the Austrian Accreditation Body relies heavily on and conforms closely not just to the EN-45000 Series, but also on internationally agreed interpretative documents such as the EURACHEM and WELAC/EAL Guides. In analytical chemistry this puts heavy emphasis on method validation as a prerequisite to accreditation.

Further information can be obtained from:

Bundesministerium für wirtschaftliche Angelegenheiten
Sektion IX/2 – Technik und Innovation
Landstraße Hauptstraße 55–57, A-1010 Vienna
Telephone + 431-71102-251 or -256 or -248; Telefax + 431-714 3582

2.3.2
The Swiss Accreditation Service

Hanspeter Ischi

A rapidly increasing infrastructure of accredited bodies is available in Switzerland providing the following types of services in almost all technical sectors:

- Calibration and verification
- testing
- certification of products, quality systems, environmental management systems and personnel
- inspection

The Swiss Accreditation Service (SAS) is the nationally recognised accreditation body in Switzerland, responsible for all types of accreditation in the mandatory as well as in the voluntary sector. As a governmental organisation, SAS is based on the Ordinance on the Swiss Accreditation System which came into force in 1991.

SAS is operated by the Swiss Federal Office of Metrology (OFMET) which launched a national recognition scheme for calibration laboratories as early as 1986.

According to the ordinance, accreditation is granted by the OFMET director based on a comprehensive assessment by SAS and after an independent evaluation of the assessment reports by a Governing Board. This Governing Board, elected by the Swiss Government, consists currently of 8 members representing the Swiss industry and economy, federal and cantonal authorities, institutes and universities.

SAS assessments are directed by its own staff. The 9 SAS leading assessors are also responsible for the assessment of the quality system of all applicants. These leading assessors have an academic degree, industrial experience of at least ten years and a broad technical education. Additionally, they are extensively trained in quality management.

For the technical part of the assessments, SAS engages technical experts with wide experience in the applicant's field who receive an additional training by SAS before they are engaged in the assessment of an applicant. In addition, the technical experts form the sectoral committees, where assessment procedures and technical requirements for given technical sectrs are discussed and defined.

The Swiss Accreditation System

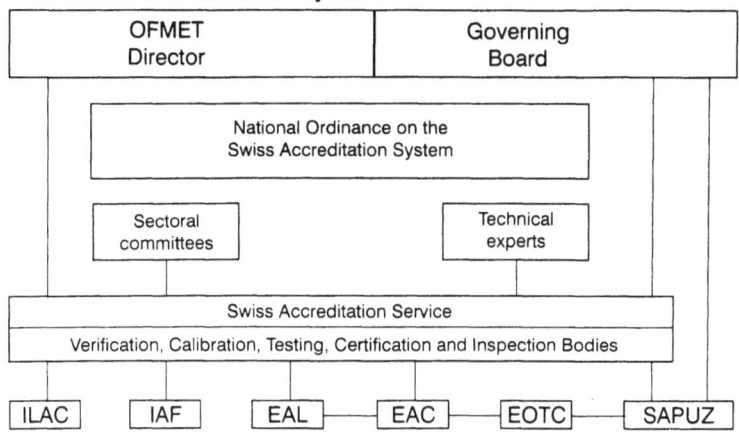

Once an organisation is accredited, it will be regularly monitored by SAS. Surveillance includes assessment of the quality system and technical aspects and is therfore conducted together with technical experts.

SAS performs its technical duties in favour of the Swiss economy. It strives to perform its duties correctly, efficiently and quickly at a high level of quality, according to fully documented procedures. SAS works therefore in accordance with all SNEN standards of the 45000 series and the corresponding ISO Guides.

The SAS accreditation procedure includes the following steps:

- Application form sent to SAS after a preliminary discussion
- Informative discussion in order to define the scope of accreditation
- Nomination of the assessment team
- Preparation of the assessment (documentation review)
- Assessment
- Evaluation of the assessment report by the Governing Board
- Accreditation granted by the director of OFMET
- Periodic surveillance of accredited bodies.

The performance of SAS is highly valued and recognised by the members of the EAL and the EAC Multilateral Agreements, both of which have been signed by SAS.

EAL is the European Cooperation for the Accreditation of (Calibration and Testing) Laboratories, EAC is the European Cooperation for the Accreditation of Certification Bodies. Signatories of both multilateral agreements are ready to promote the recognition of accredited bodies of all the other signatories in their own country.

SAS has already concluded and will also enter into further bilateral or multilateral agreements for mutual recognition with bodies outside Europe, always in accordance with EAL's and EAC's corresponding policies.

On the national level, SAS is actively cooperating with the Swiss Committee for Testing and Certification (SAPUZ, Schweizerischer Ausschuß für Prüfung und Zertifizierung) which in turn is a technical committee of the Swiss Association for Standardisation (SNV). SAPUZ was created as a mirror organization of the European Organization for Testing and Certification (EOTC), in order to create firm European links for technical harmonisation and free trade. SAS cooperates with representatives of industry and trade, government agencies, the research establishment and testing community within various technical committees of SAPUZ, such as "Calibration", "Quality assurance", "Inspection" and "Testing", the latter representing EUROLAB-CH and EURACHEM-CH.

This collaboration of all patients concerned with accreditation and certification of products and services within the transparent network of SAPUZ provides good communications and comprehensive policies.

A large number of testing laboratories in the field of chemistry have already been accredited or are in the process of accreditation. Since the responsibility for the inspection according to the rules of GLP is with other governmental authorities, a fruitful collaboration has already been set up between these bodies and SAS in order to recognise as much as possible from each other's system.

Addresses of accredited bodies together with the services they provide are published in the index of the Swiss Accreditation Service (SAS). The complete scope of each accredited body is given in an individual register.

Swiss Accreditation Service
Swiss Federal Office of Metrology
Lindenweg 50
CH-3084 Wabern
Telephone + 41 31 963 31 11
Telefax   + 41 31 963 32 10

2.3.3
The German Accreditation System

U. Böshagen

2.3.3.1
The BDI Model

Because of the dual system in Germany, the system had to be designed in a way that it would be transparent and comparable in the European setting. Therefore, for ten years, the design of the German accreditation system has been a subject that has closely occupied numerous employers' associations and institutions of commerce and industry in the follow-up of the target guidelines of the policy of the Commission of the European Union. The original idea was to place the whole accreditation system under state jurisdicton. This was basically because the accreditation systems of the European partner states are also pronouncedly governmental structures. The business associations, however, refused to abandon the strongly developed and operational private sector testing and accreditation, and, since then, have tried almalgamate the accreditation bodies organized in the private sector and the laboratories of the statutorily regulated field into a homogenous system. It has been the economy which has dictated that the State is not sole organ competent for accreditation.

Against this background, and, in order to react to developmets emanating from Brussels at the end of the 1980s, the idea of creating a confidence inspiring

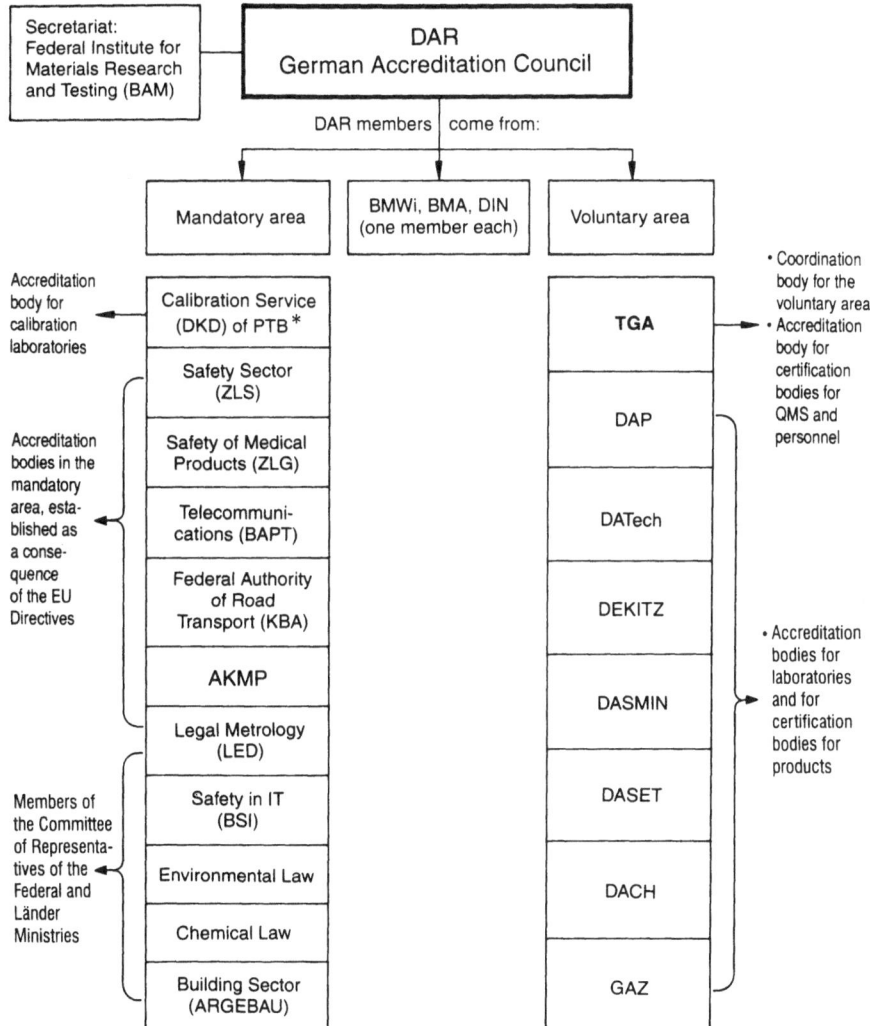

* Accreditation by governmental accreditation
  body (PTB) in the voluntary area

instrument in the private sector also emerged. The co-called "BDI" or "TGA" model, whose central idea was developed by the special BDI committee Technology and Law in November 1988, contained these elements of the German accreditation system which have been realized today: The Joint Agencies for Accreditation as the instrument and the fusion of all private sector accreditations, form part of the common German accreditation counsel in which the State also operates to the same extent but for the regulated field of accreditation. At that time, the committee recommended their preference of a pluralistic private sector accreditation body with the participation of concerned circles to a state-managed centralised institution.

In August 1990, TGA – Trägergemeinschaft für Akkreditierung GmbH (Joint Agencies for Accreditation, Co., Ltd) – was inaugurated as part of the German accreditation system that was to be set up. Founding members of the TGA are 12 leading organizations of commerce and industry themselves having no commercial interest in accreditation. The security is provided by the general committee of the TGA, where the specialist interests of industries are coordinated, also in order to foster the transparency requested throughout Europe.

This general committee coordinates sectors related to industries which grant independently accreditation to mechanical and electrical engineering, chemistry and other fields.

In addition to the general committee which acts in an honorary capacity, the management dedicates itsef to the day to day work. Corresponding to the requirements of the underlying standards, the work is supported by further necessary committees, i.e. the arbitration committee, the structure and surveillance committee as well as the shareholders' meeting or its supervisory board. Therefore accreditation in the non-regulated area is only possible if the accreditation body is part of the total system of TGA.

In conclusion one should mention again the main tasks of the TGA and its general committee within the system:

- to coordinate the activity of the accreditation bodies within the system of the TGA, for example, to coordinate the compilation of technical requirements for accreditation,
- to hammer out policies for the accreditation bodies as well as to design uniform accreditation documents,
- to appoint the president of the conciliation committee and the court of arbitration of the TGA,
- to organize the national and international representation of the TGA, when general questions of accreditation have to be answered,
- to assess, through supervision by the structure and surveillance committee, the correct working of accreditation bodies, as well as
- to coordinate the representation of accreditation bodies in the German accreditation counsel.

The essential sectors that have given rise to accreditation bodies are among others: the Deutsche Akkreditierungssystem Prüfwesen, the Deutsche Akkreditierungsstelle Technik, the Deutsche Koordinierungsstelle für die IT-Normenkonformitätsprüfungen und -zertifizierungen, as well as the Deutsche Akkreditierungsstelle Chemie (German Accreditation Body Chemistry).

Though the TGA essentially coordinates the independant sectorial accreditation work, it also grants accreditation in the case of quality management and personnel resistrars.

Therefore, the TGA is one pillar of the DAR, the second essential pillar being constituted by the accreditation authorities of the regulated field. Permanent collaborators within the DAR are, in addition to those just mentioned, representatives of the Federal Ministry of Economic Affaires (Bundesministerium für Wirtschaft), the Federal Ministry for Employment and Social Affairs (Bundesministerium für Arbeit und Sozialordnung) and the German Institute for Standardization r.a. (Deutsches Institut für Normung e.V.).

The DAR has two essential tasks:

The coordination of both provinces, the statutorily regulated and the non-regulated field, as well as ensuring transparency of the whole system to the outside world, i.e. especially to Brussels and to the Common Market. This includes in detail:

1. to coordinate activities carried out in Germany in the field of accreditation and recognition of testing laboratories, certification and surveillance bodies,
2. to safeguard German interests in national, European, and international institutions which are concerned with general questions of accreditation respective recognition, and
3. to keep a central German register for accreditation and recognition.

The DAR acts as a working group supported by the Federal Government, the governments of the individual states and by German commerce and industry. It has no legal status. Its office is run by the Bundesanstalt für Materialforschung und -prüfung. According to the Global Concept for Testing and Certification, accreditation means an intensification of investigating the question of whether an authorized laboratory has carried out its work on the basis of the correct standards and directves. As a general rule, this eliminates retesting.

2.3.3.2
Cooperation Between Regulated and Non-regulated Area

How far does work between both areas go beyond their cooperation within the German Accreditation Counsel? It can be stated that in the sectors related to industries and in their steering committees, there is already close and successful cooperation of representatives of both the federal accreditation authorities and the nonregulated area. Technical processes are not influenced by whether the impetus for certification or accreditation is given by law or by a regulation, or, if it is initiated by the manufacturer's voluntary request. Therefore, since the very beginning, the technical cooperation between both areas has been more effective and profound than one would have expected from the fundamental considerations for the implementation of the dual system.

Addressess:

DAR – Deutscher Akkreditierungsrat
c/o Bundesanstalt für Materialforschung
und -prüfung
Unter den Eichen 87
D-12203 Berlin
Germany
Telephone +49-30-8104-1942
Telefax   +49-30-8104-1947

TGA-Trägergemeinschaft für
Akkreditierung GmbH
Stresemannallee 15
D-60596 Frankfurt
Germany
Telephone +49-69-63009-111
Telefax   +49-69-63009-144

DACH – Deutsche Akkreditierungsstelle
Chemie GmbH
Stresemannallee 15
D-60596 Frankfurt
Germany
Telephone +49-69-630003-20
Telefax   +49-69-630003-23

DAP – Deutsches Akkreditiersystem Prüfwesen
GmbH
Rudower Chaussee 5, Geb. 13.7
D-12489 Berlin
Germany
Telephone +49-30-67059110/20
Telefax   +49-30-67059115

DASMIN – Deutsches Akkreditier-
system Mineralöl
Kapstadtring 2
D-22297 Hamburg
Telephone +49-40-63900471/72
Telefax   +49-40-63900736

2.3.4
The Danish Accreditation System, DANAK

H. Jensen

The Danish Accreditation System, DANAK, is the oldest in Europe and was established back in 1973 as "States Tekniske Prøvenævn" (STP) (the National Testing Board) and the only systems in the world which are older than the Danish are the ones in Australia and New Zealand.

DANAK is run by the Danish government through the Danish Agency for Development of Trade and Industry (EFS). The Danish Accreditation is voluntary and open to all qualified laboratories in the private as well as the public sector.

Normally, accreditation of a testing laboratory is granted for a five-year period. During the period DANAK will carry out three surveillance visits to verify that the laboratory is continuously meeting the requirements for accreditation.

If a laboratory applies for renewal after the five-year period, a renewal visit will be conducted.

At the beginning DANAK consisted of only accrediting testing and calibration laboratories but today this is only one of DANAK's four key areas.

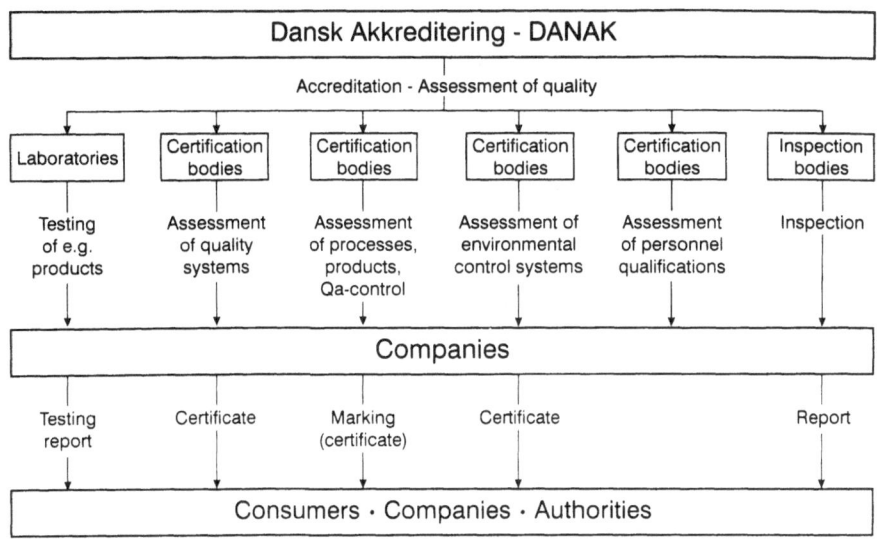

The other three areas are:

– accreditation of bodies for certification of products, systems and staff
– accreditation for inspection
– accreditation of bodies to certification of environmental control systems, companies and persons carrying our verification of environmental matters. (See Figure).

DANAK also carries out GLP-inspection of laboratories in areas which are regulated by the ministry for the Environment. The GLP-inspection is carried out according to OECD's rules.

DANAK has signed both of the multilateral agreements within EAL, that is both the multilateral agreement within the former WELAC (testing) and the one within the former WECC (calibration).

2.3.4.1
Structure of DANAK in the Area Concerned with Accreditation of Testing and Calibration Laboratories

The head of DANAK is the head of the department "International Trade Conditions" in EFS. The lead assessors are all employed by EFS while the technical assessors are experts from universities and industry, who are employed on a contractual basis by EFS.

DANAK has an Advisory committee which is consulted in the case of fundamental questions concerning the principles of accrediting laboratories. This committee has to be heard concerning all requirements which are going to be placed on the accredited laboratories. The members of the committee are representatives from other ministries using DANAK to monitor laboratories working for them, representatives from the accredited laboratories and representatives from industry.

In addition to the advisory committee there are some sectoral committees working with topics specific to a special type of laboratory. At present there are 3 committees which are concerned with calibration, environmental analysis and food analysis. The members of the sectoral committee are experts – often technical assessors within the area, representatives of ministries concerned with the area, representatives of the users of accredited results and representatives of laboratories accredited within the area.

It will normally be these committees, which decide if there is a need for extra technical requirements to be set up in an area, in order to specify which qualifications a laboratory in this specific field needs to fulfil to comply with EN 45001. A lead assessor will then be put in charge of a group which will prepare the requirements.

After discussion in the sectoral committee, the proposal is sent on to the advisory committee for hearing.

The sectoral committees also recommend to DANAK which proficiency tests DANAK should carry out.

2.3.4.2
Accreditation of Chemical Laboratories

Accreditation of testing and calibration laboratories is carried out according to EN 45001–3 and additional technical requirements.

Only one of these technical requirements concerns chemical laboratories and it sets up some specific requirements for the personnel allowed to sign accredited reports which contain results of microbiological tests.

All laboratories which are accredited to do chemical analysis have to participate in a given proficiency test within the area where the laboratory is accredited, if DANAK so requires and this at the laboratory's expense.

There are two sectoral committees working with matters concerning chemical laboratories. These committees work with environmental analysis and food analysis.

There are today (March 1995) about 70 laboratories in Denmark which have been accredited to do chemical analysis and/or microbiological tests and that is about 40 % of the total number of accredited laboratories in Denmark.

In Denmark, one laboratory has been accredited to do chemical analysis in a mobile laboratory, that is a laboratory installed in vehicle which can work under field conditions and still fulfil the requirements in EN 45001.

This laboratory has also been accredited to carry out nitrateanalysis on a small spectrophotometer which has been build into a suitcase and which therefore also can be used in the field.

Seven laboratories have been found to comply with the GLP rules of OECD.

Address:

DANAK
Erhvervsfremme Styrelsen
Tagensvej 137
DK-2200 Copenhagen N
Denmark
Telephone + 45 35 86 86 86
Telefax   + 45 35 86 86 87

2.3.5
Finas – The Finnish Accreditation Service – Laboratory Accreditation

Christina Carlsson

2.3.5.1
General

The national accreditation body in Finland is the Centre for Metrology and Accreditation. Laboratory accreditation (authorization) activity in Finland began in 1978, when the Technical Inspection Centre authorized the first calibration laboratories. The accreditation activities in Finland were reorganized in the spring of 1991 when the Centre for Metrology and Accreditation was established. The new Centre, which is a governmental body with public funding, is the national accreditation body according to the Government Decree 1568/91 on the accreditation of bodies performing testing, inspection, certification, and calibration. The unit of the Centre for Metrology and Accreditation responsible for the accreditation activities is called FINAS (Finnish Accreditation Service).

The aim of the accreditation activity and the decision-making process has been described in the Decree (1568/91) and its amendment (1533/94). FINAS assesses the laboratories for compliance using the criteria of standard SFS-EN 45002. The laboratories have to fulfil the criteria of the standard SFS-EN 45001 and ISO/IEC Guide 25 and the calibration laboratories furthermore the criteria specified in the Additional Requirements for Calibration Laboratories. These additional requirements are based on the WECC requirements for calibration laboratories. The accreditation body can be the virtue of the decrees withdraw or suspend accreditation if the laboratory does not continuously comply with the accreditation requirements. To ensure that the criteria are fulfilled, the FINAS makes surveillance visits to the laboratories at intervals of six to twelve months. At surveillance visits both the lead assessor and technical assessors are present. The accreditation of certification and inspection bodies has similar procedures, which are not described in this paper.

2.3.5.2
Operation

The accreditation body, which operates according to the SFS-EN 45003 standard and its Quality Manual and procedures, is run by the head of the accreditation service assisted by four full-time assessors, who all have been trained also for quality system assessment. Technical experts from different fields of technology have been trained to act as technical assessors. The accreditation body arranges three-day training courses for the assessors. The accreditation body has now registered about 120 technical assessors.

FINAS accredits all kinds of laboratories that meet the requirements in SFS-EN 45001 and ISO/IEC Guide 25. The applicant laboratories are assessed according to SFS-EN 45002. FINAS has accredited 31 testing laboratories and 29 calibration laboratories (end of March 1995) and made surveillance visits to the accredited laboratories. The chemical laboratories represent half of the accredited testing laboratories. The head of the accreditation body nominates a lead assessor to be responsible for the handling of an application. The lead assessor selects a competent team, so that the scope of accreditation in the application is covered. The assessor team is formally appointed when the laboratory agrees to the members of the team in writing. The lead assessor presents an estimation of the total costs of the assessment, which the laboratory is asked to sign before the assessment is started. The assessment visit will be planned and agreed on with the assessment team and the applicant laboratory. Special attention is placed on results from interlaboratory comparisons, where available. When no results from such comparisons are available or not considered to be reliable, the laboratory will be asked to perform proficiency tests during the visit. At the end of the assessment visit the findings are reported to the laboratory and corrective actions are agreed on in writing. The lead assessor will give a summary of the results and indicate the team's recommendation to the accreditation body. Each assessor will make a report on his observations. When the corrective actions are performed and reported, the assessors will give a statement on the results of the corrective actions and their recommendation to the accreditation body. The lead assessor writes a concluding report and a draft accreditation decision.

The accreditation decision is then issued by the FINAS. Accreditation is granted for a limited time, usually 4 years. The laboratory must comply continuously with the accreditation requirements. If these requirements are not met the accreditation body can withdraw or suspend accreditation or reduce the scope of accreditation. The laboratory is expected to inform FINAS of any relevant changes that influence the accredited laboratory. If the legal status of the accredited laboratory changes, e.g. with a change of the ownership, its effect on the laboratory's activities must be monitored and a new decision on accreditation has to be made. The scope of accreditation can, at the laboratory's request, be changed or enlarged. The accredited laboratories may use the FINAS logo. The conditions for the use of the FINAS logo or reference to FINAS are set out in the accreditation decision.

2.3.5.3
Information

The accreditation body issues interpretation documents for applicant laboratories on how to comply with the accreditation criteria. It also maintains records on accredited calibration and testing laboratories, and monitors and controls their operation compliance with requirements. The accreditation body also issues a directory of the accredited laboratories, containing contact information and scope of accreditation. The accreditation body organizes training required for the relevant operations, carries out publishing and information activities in the field, and takes part in arrangements and development of national and international interlaboratory comparisons. It follows the general development in the field of accreditation, proposes measures for further development of the relevant procedures, participates into EAL, ILAC activitis, and concludes agreements on mutual recognition of the accreditation of laboratories with corresponding foreign organizations within EAL.

Christina Carlsson
Centre for Meteorology and Accreditation
FINAS
P.O. Box 239
00181 Helsinki
Telephone +358061671
Telefax   +35806167341

2.3.6
The Italian Accreditation System

M. Grazia Del Monte

Since 1985 there has been a tendency in Europe towards the adoption of regulations and voluntary certification in order to bring about the free circulation of products within the Single Market. The result has been a series of initiatives which have progressively led to the construction of a European System aimed at harmonising tests and certification.

Italy has had a long tradition of technical specifications regulated internally by laws. The only real certification of consequence which existed was the application of these laws in testing or certifying products.

Public Administration checked compliance with technical obligatory rules. Episodes of voluntary certification also occurred in compliance with regulations.

Therefore, in order to bring institutions into line with the European system, no radical changes were necessary.

Italy, in compliance with the EEC directives, has set up a voluntary structure able to ascertain the technical competence of the laboratories in which the tests and qualification measurements are performed as well as the quality system of the industrial firms. An important element of this structure are the Accredita-

Italian quality system structure

Name	Type of administration	Competence	
Ministries and Technical State Organisations	Public Administration	Issue of mandatory technical regulations	Public Administration CNR
UNI	Association between private subjects and Public Administration	Issue of voluntary standardisation	UNI
CEI	Association between private subjects and Public Administration	Issue of voluntary standardisation (electric and electronic sectors)	CEI
SINCERT	Association between private subjects and Public Administration established by UNI and CEI	Accreditation of certification agencies. Mutual recognition with similar organisations	SINCERT
SINAL	Association between private and Public Administration, established by UNI and CEI	Accreditation of test laboratories. Mutual recognition with similar organisations	SINAL
SNT	Metrology institute and calibration centres	Preservation and spread of national samples. Calibration of the instruments	SNT
CNMR	Association between private subjects and Public Administration established by Centre Sviluppo Materiali	Metrology in chemistry and control of the national activities on certified reference materials	CNMR
CERTIFICATION Agencies	SPA, SRL, Associations, etc.	Certification the quality systems	Certifying Bodies
TEST-LABORATORIES	University, Research Centres, SPA, SRL, etc.	Test and analysis of products	Testing Laboratories
CONTROL Organisations*	Technical State Bodies	Controls on the market	Controlling Bodies

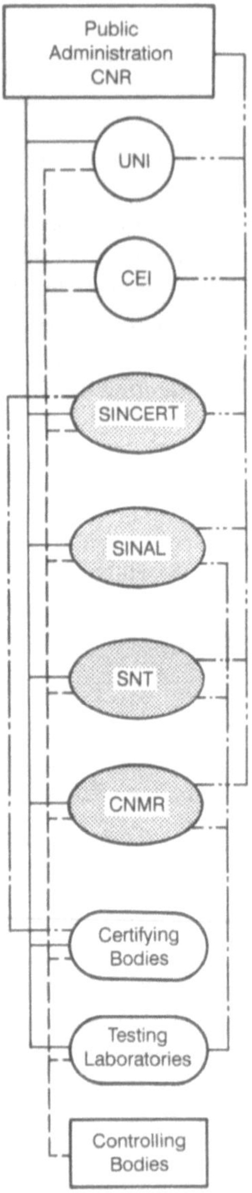

* to be founded

— - — Coordination   — — — Technical Stds.
———————— Technical Rules   —— · —— Accreditation

Fig. 5

tion and Certification Bodies which, acting as independent bodies with all parties participating, guarantee to the users the competence and the impartiality of the testing laboratories when carrying out tests and the uniformity of quality of the Certificate apparatus.

This voluntary structure which together with Public Administration constitutes the "Italian Quality System" (Fig. 5), which includes, besides the accreditation bodies SINAL (National System for the accreditation of laboratories), SINCERT (National System for the accreditation of the certification Organisations) and the Certification Agencies, the standardisation bodies UNI (National Italian Association of Standardization) and CEI, (Italian Electrotechnical Commission), the agencies for calibration, (SNT – National System of Calibration), and reference materials, (CNMR – National Centre for Reference Materials).

The task of UNI and CEI is to produce the voluntary standardisation regulations.

SNT, devised for the accrediting of calibration laboratories, deals with physical metrology and verifies the validity of the calibration samples used in the tests when quality assurance requirements are applied.

CNMR established in 1991, deals with metrology in chemistry, diffuses information on Reference Materials, controls the national production of reference materials and verifies, in Italy, the validity of every type of reference material for accredited and certified tests.

SINAL deals with the accreditation of testing laboratories in every field of application (chemical, mechanical, food, environmental, biological, etc.). It verifies and checks periodically the conformity of the testing laboratories to the standard UNI CEI EN 45001.

SINAL was established on April 26, 1988, by UNI and CEI, under the auspices of the Ministry for Industry, of Commerce and Art and Craft, of CNR (National Council of Research), of ENEA (Authority for new Technology concerning Energy and the Environment), of the Chambers of Commerce, Industry, Agriculture, and Arts and Crafts. It is a non-profit making association, legally recognised by the 9/9/91 Act, passed by the Ministry of Industry, Commerce, and Arts and Crafts. SINAL headquarter is located in Rome, Via Campania 31.

The aim of the association, is to grant Accreditation to Italian and foreign testing laboratories verifying and guaranteeing that they conform to UNI CEI EN 45001 and to SINAL criteria.

By promoting mutual recognition and/or reciprocity agreements with analogous organisations in other countries, SINAL contributes to the reduction of technical barriers to international trade. On June 14, 1993 SINAL signed the mutual recognition agreement with the accreditation bodies of Australia, Denmark, Finland, France, Germany, Hong Kong, Ireland, Netherlands, New Zealand, Norway, Spain, Sweden, Switzerland, and the United Kingdom.

This agreement states the equivalence of the operation of the above-mentioned accreditation bodies; on the basis of this agreement each signatory will:

– recognise the other accreditation bodies that are signatories to the agreement as equivalent to itself

- recommend acceptance on an equal basis with those of its own accredited laboratories of the test reports from the laboratories accredited by the other accreditation bodies that are signatories to the agreement
- promote the acceptance of test reports of laboratories accredited by the other accreditation bodies that are signatories to the agreement by all the users in its own country
- investigate all complaints initiated by a signatory to the agreement resulting from test reports issued by its own accredited laboratories.

SINAL grants, maintains or withdraws Accreditation and takes the necessary actions of surveillance with the aim of guaranteeing the competence and impartiality of the laboratory.

In particular,

- SINAL assesses and oversees the technical and organisational requisites of testing laboratories in order that metrological references, reliability and repeatability of the procedures adopted, the use of adequate instrumentation, the professional competence of personnel are guaranteed and conform to UNI CEI EN 45001 and SINAL criteria and/or those in place in the Community either at the international or national level;
- on the grounds of information and results received from the evaluation visits, the SINAL board of directors, after having consulted the Central Technical Committee, deliberates the Accreditation. A certificate of Accreditation is issued to the laboratory with the accreditation number and the list of test for which the laboratory has been accredited. All information acquired by SINAL concerning accreditation remain confidential;
- after having obtained accreditation, the laboratory will receive a first surveillance visit after 6 months, then surveillance visits at yearly intervals and a full evaluation visit every 4 years to renew the accreditation. During the period in which the laboratories are covered for accreditation, they may be asked by SINAL to take part in proficiency tests.
- SINAL publishes the accreditation procedures, the list of the accredited laboratories, and the eventual withdrawal or suspension of accredited laboratories on the national official journal "U & C".
- SINAL cultivates relations with analogous bodies of other countries, with the aim of drawing up mutual recognition agreements,
- participates in the work of international and/or supranational bodies working in the field of testing laboratory accreditation,
- collaborates with national standardisation bodies with the aim of improving or updating technical standards,
- promotes studies, meetings, and initiatives in its area of specific professional competence, with associations and bodies at national and international level.

About one hundred laboratories have obtained SINAL accreditation. A few University laboratories are now starting, to ask for SINAL accreditation.

At present SINAL is studying procedures of agreement with some Ministries, such as the Ministry of Health, in order to avoid duplication of work in the evaluation activity for accrediting, for instance, testing methods in the field of their specific competence.

SINCERT is a non-profit-making organisation, that deals with the accreditation of certification bodies of quality systems, products, staff. It verifies and checks periodically the conformity of the certifying organisations to the standard UNI CEI EN 45011-12-13.

SINCERT was established in November 1991, by UNI and CEI, under the auspices of the Ministry for Industry, of Commerce, and Arts and Crafts, of CNR, of ENEA, of Ministries of the Environment, Foreign trade, Defence, Internal affairs, Public works, Posts and Telecommunications, Transport, of the Chambers of Commerce, Industry, Agriculture, and Arts and Crafts, of category associations. SINCERT headquarters is located in Milan, Piazza Diaz 2.

The aim of the association, is to strengthen, through accreditation, authority and reliability of bodies operating in the field of certification, both at Italia and international level.

SINCERT verifies procedures adopted by certifying bodies. SINCERT assessors, in fact, take part in one of the inspection visits to a firm to be certified. After having obtained accreditation the certifying body will receive surveillance visits at intervals, according to the decision of Accreditation Committee of SINCERT.

SINCERT is member of EAC (European Accreditation Council, the organisation working towards the mutual recognition of the accreditation agencies).

Certification bodies grant certification to industries according to EN ISO 9000 standards.

In Italy, they are organised in a quite different way from the scheme adopted in Europe.

Europe Certification bodies are devoted to all the technological sectors, while in Italy each certification body is related to a particular industrial sector. Among these bodies, we can list CERTICHIM, (Certification of Chemistry sector), IGQ (Institute for quality assurance in the surgical field), IIP (Italian Institute for plastic), ICIM (Institute of industrial certification for mechanics), IMQ (Italian Institute for quality), etc.

Addresses:

UNI Ente Nazionale Italiano di Unificazione
Via Battistotti Sassi 11
20133 Milano
Italia
Telephone +39 2700241
Telefax   +39 670106106

CEI Comitato Elettrotecnico Italiano
Viale Monza 259
20126 Milano
Italia
Telephone +39 225773.1
Telefax   +39 225773.222

SINAL Sistema Nazionale per
l'Accreditamento di Laboratori
Via Campania 31
00187 Roma
Italia
Telephone +39 64871141/76
Telefax   +39 64814563

SINCERT Sistema Nazionale per
l'accreditamento degli organismi di
Certificazione
Piazza Diaz 2
20123 Milano
Italia
Telephone +39 286464374
Telefax   +39 272023085

SNT Sistema Nazionale di Taratura
Strada delle cacce 91
10135 Torino
Italia
Telephone +39 11 3488933
Telefax   +39 11 346384

CNMR Centro Nazionale per i Materiali di
Riferimento
c/o Centro Sviluppo Materiali
Via di Castel Romano 100
00129 Roma
Italia
Telephone +39 6 5924085
Telefax   +39 6 5050250

2.3.7
The Irish National Accreditation Board

F. T. Smyth

2.3.7.1
Background

Accreditation in Ireland started in 1985 with the formation of the Irish Laboratory Accreditation Board as part of the National Standards Authority of Ireland (NSAI). When it was first set up it was responsible for accreditation of testing and calibration laboratories.

In 1992, it became a signatory to the Western European Calibration Co-operation (WECC) and Western European Laboratory Accreditation Co-operation (WELAC) multilateral mutual recognition agreements.

In 1994 it was reorganised, renamed as the Irish National Accreditation Board and given responsibility also for the accreditation of certification bodies. It is now a division of Forfás, the policy and advisory board for industrial development in Ireland. When WECC and WELAC merged in 1994 to form the European Co-operation for Accreditation of Laboratories (EAL) the Irish National Accreditation Board became a signatory to the EAL multilateral agreements on both testing and calibration.

EAL has now signed bi-lateral mutual recognition agreements with the national accreditation bodies of Hong Kong, Australia, New Zealand and South Africa.

2.3.7.2
Scope of Operations

The Irish National Accreditation Board is responsible for the following activities:

- National accreditation scheme for testing and calibration laboratories in accordance with harmonised European standards EN 45001, EN 45002 and EN 45003.
- National Monitoring Authority for GLP (Good Laboratory Practice) in Ireland in accordance with Statutory Instrument No. 4 of 1992; European Communities (Good Laboratory Practice) Regulations, 1991.
- National accreditation scheme for certification bodies in accordance with the harmonised European standards EN 45011, EN 45012, and EN 45013.

Organisation diagram of the Irish Accreditation System

- Accreditation of Environmental Verifiers under the EC Eco-Management and Audit Scheme.
- "Competent Body" (Registration Body) for the EC Eco-Management and Audit Scheme.
- National representative on EU Commission, EMAS Regultory Committee
- National representative on the OECD GLP Panel
- National representative on EAL, the European Co-operation of Accreditation of Laboratories
- National representative on EAC, the European Accreditation of Certification Bodies

2.3.7.3
Statistics

Total number of laboratories accredited to date:	51
These comprise:	
Calibration laboratories	10
Chemical laboratories	20
Non-destructive testing laboratories	3
Microbiological laboratories	9
Mechanical testing laboratories	4
Personnel dosimetry laboratories	1
Electrical laboratories	3
Medical laboratories	1

Address:
The Irish National Accreditation Board,
35–39 Shelbourne Road,
Ballsbridge,
Dublin 4,
Ireland

2.3.8
The National Accreditation Body of Norway

Khalid Saeèd

2.3.8.1
General History of the Norwegian Accreditation System

Norwegian Accreditation (NA) is a part of Norwegian infrastructure in the general field of measurement, the National Measurement Service (NMS). The NMS was established by the Norwegian "Law on weights and measures" with of 1875 and operates under the Ministry of Industry and Energy.

NMS was given the responsibility for accreditation of calibration laboratories in 1987 which was denoted by Norwegian Calibration Service (Norsk Kalibreringstjeneste) for historical reasons.

In 1990, Norwegian Accreditation was set up as one of the three technical sections of NMS in accordance with Proposition no. 106, 89/90 to the Norwegian Parliament on accreditation in Norway. Based upon the same Proposition, the Royal Decree in June 1991 placed the responsibility of accreditation of certification and inspection bodies in addition to test and calibration laboratories and in October 1993 GLP-compliance on NA. Accreditation of environmental testers according to the EMAS-regulation was added in 1994.

The two other technical sections being; Legal Metrology and the National Standards Laboratory.

2.3.8.2
Organisation

Norwegian Accreditation is led by a Section Director who report to the General Director of the NMS.

The Ministry of Energy and Industry has appointed a Technical Advisory Board (Fagstyre) whose task is to ensure that accreditation and GLP-compliance services are estabished and developed in accordance with the recognised international principles and the national needs.

The organisational chart of Norwegian Accreditation is presented in Fig. 6.

For laboratory accreditation which includes testing and calibration, NA has its own staff of technical officers who are responsible for the conduct of assessment and surveillance visits of the laboratories in accordance with the procedures approved by EAL (European Co-operation for Accreditation of Laboratories). See section 2.3.8.5 for the conduct of technical assessment.

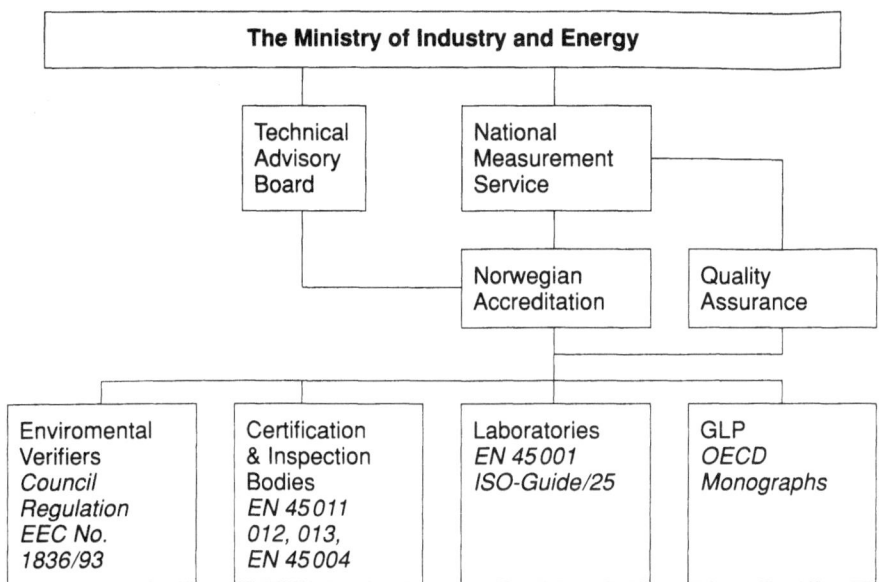

Fig. 6. Organisational Chart

2.3.8.3
Tasks for Norwegian Accreditation

The major tasks for NA are the following:

- to provide accreditation services to Norwegian laboratories in accordance with international standards based upon EN 45001 and ISO guide 25.
- to provide a GLP-compliance monitoring system for testing of chemicals including drugs, pesticides, industrial chemicals etc. in accordance with OECD's GLP-principles.
- to provide accreditation services to certifying enterprises according to EN 45011, EN 45012, EN 45013 and EN 45004.
- to perform accreditation of environmental testers in accordance with the Council Regulation (EEC) No 1836/93 of 29 June 1993.
- to provide information about accreditation and the accredited enterprises.
- to ensure that the certificates issued by the Norwegian laboratories accredited by NA have international acceptance.
- to safeguard Norwegian interests in international cooperation on accreditation.

2.3.8.4
Multilateral Agreements (MLA)

NA is a signatory to multilateral agreements established by EAL (European cooperation for Accreditation of Laboratories) on calibration and test laboratories, and EAC (European Accreditation Co-operation) on accreditation of certification bodies.

2.3.8.5
Application and Assessment Procedure for Laboratory Accreditation

After a formal registration and acceptance of an application, the technical officer conducts a pre-assessment visit to the laboratory either alone or with the technical assessor. At this stage, the laboratory's quality system is evaluated in relation to the standard. A report is produced which points out the areas to be improved before the main assessment can be conducted. The laboratory is required to have carried out an internal audit of the quality system prior to conducting the main assessment. During the main assessment, the objective observations are recorded on non-compliance. Each non-compliance is designated category 1, 2 or 3 depending upon the severity of non-compliance. Serious non-compliance that poses a threat to the quality system is categorised in category 1, while a minor non-compliance is categorised as category 3. Several non-compliances within one and the same area is categorised in category 2. In the concluding session of the main assessment, the lead assessor presents a summary report and his/her recommendation regarding accreditation. Each technical assessor writes a report based upon his/her findings during assessment. The lead assessor writes a report which contains his/her own findings in addition to findings of the technical assessor(s). This report is made available to the laboratory for their comments regarding factual information. The technical office produces an accreditation report and presents his/her recommendation for accreditation to the Section Director of NA. The General Director of NMS issues an accreditation certificate after recomendation from the Section Director.

2.3.8.6
Assessment Team

The main assessment of the applicant is conducted by an assessment team consisting of a lead assessor and at least one technical assessor. For laboratories applying for multi scope accreditation in different technical fields, more than one technical assessor may be included in the assessment team. Assessors are normally external consultants trained by NA to perform technical assessment. Appointment of external assessment personnel is made through a written contract after formal acceptance of the assessment team by the applicant. The assessor is normally contracted for a specified task.

NA has approximately 80 qualified Norwegian external technical assessors who cover various technical fields and who are contracted more or less on regular basis.

Requirements for assessors:

Lead assessor: A lead assessor, an internal or external person, is appointed to assess an organisation in accordance with the requirements laid down in EN-45002. It is NA's requirement that a lead assessor must have participated in an assessor training course and at least four assessments, two as an observer and two as lead assessor under the supervision of a qualified lead assessor.

Technical assessor: A technical assessor, an internal or external person, is appointed to perform a technical assessment of an organisation in accordance with the requirements laid down in EN-45 002. It is NA's requirement that a technical assessor must have participated in an assessor training course and if possible, at least one assessment as an observer before he and she is employed as an assessor.

2.3.8.7
Collaboration with Other Accreditation Bodies

It is the policy of NA to contract qualified external assessors who live in Norway and are finent in Norwegian. In some cases, however, foreign technical experts may be engaged preferably through a collaboration with other accreditation bodies who are signatories of the relevant MLA such as DANAK in Denmark, SWEDAC in Sweden and NAMAS in U.K.

2.3.8.8
Accreditation for the Regulated and Non-Regulated Area

Laboratory accreditation in Norway is in most cases voluntary (non-regulated area). However, in some fields, the appropriate law enforcing authorities such as the Norwegian Petroleum Directorate (Oljedirektoratet) and The Norwegian Pollution Control Authority (SFT) have required laboratories to be accredited in accordance with EN 45 001.

In the regulated area, accreditation is, in most cases, required for an institution that is notified in relation to EU-directives.

2.3.8.9
Sector Committees

Sector committees have been created to assist in the establishment of new fields such as GLP, environmental verifier, accreditation of notification bodies, and in important areas of established fields such as non-destructive testing, certified reference materials, computers etc. In addition, the Technical Advisory Board will take up broader technical questions and problems.

2.3.8.10
Concluding Remarks

The Accreditation Service is a relatively new establishment in Norway. There has been however, a rapid increase of interest and demand for test and calibration laboratory accreditation. Until now NA has accredited 33 test and 12 calibration laboratories covering a vast variety of different fields and approximately 50 application are being processed. The potential number of laboratories seeking accreditation in Norway is estimated to be 200 – 300.

Address:
Norwegian Accreditation
National Measurement Services
P.O.Box 6832 St. Olavs Plass
N-0130 Oslo
Telephone +47-2220-0226
Telefax   +47-2220-7772

2.3.9
The Netherlands Accreditation System

P. van de Leemput

2.3.9.1
Introduction

The Laboratory Accreditation Board of the Netherlands was established in 1986 ad has operated since that time under the shortened name STERLAB. In 1990 the activities of the foundation expanded with the accreditation of inspection-bodies; the department concerned operates under the name STERIN.

Since January 1, 1993, STERLAB and STERIN have been merged with the Netherlands Calibration Service (NKO). Until this date the NKO, founded in 1975, was an activity of the Netherlands Measurement Institute; this institution was founded by the Dienst van het Ukwezen.

The name of the foundation is now: "Dutch Accreditation Board for Calibration Laboratories, Test Laboratories and Inspection Bodies", hereafter referred to as "the Accreditation Board". The Accreditation Board operates from the Business Centre Rotterdam under the names NKO, STERIN and STERLAB.

The purpose of the Accreditation Board is:

a) to act in the Netherlands as the national accreditation body and administrator of an accreditation system for laboratories and inspection-bodies;
b) to improve the quality of calibration laboratories, test laboratories and inspection-bodies;
c) to promote, nationally and internationally, knowledge and general acceptance of the accreditation system administered by it;
d) to improve the quality of metrology in the Netherlands as applied by industry, science, inspection institutions and government;
e) to guarantee traceability in agreement with the Netherlands Measurement Institute;
f) to carry out the tasks delegated by the government in accordance with governmet supervision for calibration laboratories, test laboratories and inspection-bodies situated in the Netherlands.

The Accreditation Board's governing bodies are the Executive Board and the Supervisory Board. The Executive Board is charged with the control of the Accreditation Board and is the sole body empowered to grant accreditation.

The Supervisory Board appoints the Executive Board.

The Supervisory Board's task is to supervise the policy of the Executive Board. In addition, for several decisions, the Executive Board requires the approval of the Supervisory Board. For example decisions regarding: accreditation criteria; standard accreditation agreements; entering into international collaborative agreements; appointing Committees of Experts. The last, but not least, important task of the Supervisory Board is the approval of the Accreditation Board's annual accounts.

In addition of these administrative organs, the Accreditation Board has Committees of Experts for the individual specialist areas and a Board of Appeal.

2.3.9.2
Objectives and Accreditations

As stated above, the Accreditation Board functions in the Netherlands as the national body for accrediting laboratories and inspection-bodies, hereafter referred to as "*Institution*" or "*Institutions*". At the same time the Accreditation Board must promote the international acceptance of the Dutch system of accreditation and of accreditations granted on the basis of this system.

All this implies that the assessments must not just be good in an absolute sence, but must take place on the basis of generally accepted standards and criteria.

The criteria to which laboratories have to satisfy to receive an accreditation-certificate by the Accreditation Board are formulated in EN 45 001 and ISO/IEC Guide 25. Interpretations of these standards are given in the following documents:

- SC00, Common Criteria NKO/STERIN/STERLAB;
- SC01, Supplementary Criteria for Test Laboratories;
- SC03, Supplementaty Criteria for Research and Development;
- SC05, Supplementary Criteria for Calibration Laboratories;
- SC09, Supplementary Criteria for Veterinary Diagnostic Pathology.

The criteria for laboratories cover the relevant criteria from ISO 9001 and ISO 9002.

Where necessary and after approval by the Executive Board, Committees in Experts (in particular Coordinating Committees of Experts, Committee of Lead Assessors, Technical Committees and Working Groups) interprets criteria or creates supplementary criteria which are also to be used in assessments.

It is therefore of great importance that the institutions which apply for accreditation, supply the Accreditation Board with comprehensive information about the fields for which they want to obtain accreditation at an early stage, so that if there are no specific criteria for these fields, these can still be established by experts when necessary.

The accreditations are granted for the quality system and for specifically defined transactions of the Accreditation Board on account of the European and international rules. Mobile Laboratories are also eligible for accreditation.

The STERLAB Accreditation for research-laboratories is based on NEN 3417 and the interpretation with it by the Coordinating Committee of Experts is laid down in document SC03, "Supplementary Criteria for Research Laboratories".

To achieve the above, a number of regulations have been established in which the procedures for accreditation and the assessments which have to be made for this purpose are laid down, together with the rights and duties of the Accreditation Board as regards the accredited institution, and vice versa.

The Accreditation Board has created the possibility to draw up "accreditation programmes" which cover specific criteria to which laboratories have to comply in order for them to be accredited for that specific accreditation programme.

An accreditation programme in the simplest form may contain a set of facilities which have to be a minimum part of the accreditation.

An accreditation programme may also include specific demands laid down by for example the Government if the laboratory wants to carry out tasks for the Government. This is possible, especially for Notified Bodies.

Accreditation Programmes are not a substitute for accreditation by EN 45001, but form an easily recognizable part of the scope of accreditation for the market.

Accreditation programmes are drawn up by working groups of interested parties set up by NKO/STERIN/STERLAB. A legislation may also determine the contents of a programme. An institution may of course be accredited for parts of the accreditation programme, but they may not use the name of the accreditation programme.

On 1 May 1995 there are about 120 accredited testing laboratories and 68 calibration laboratories.

Through Videotex (06-7300) with login name STERLAB, the Accreditation Board publishes yearly overviews of the accredited laboratories and inspection-bodies. The overviews are drawn up in Dutch and in English.

A short description of the NKO laboratories is given in the yearly "Short-Form Directory". A more detailed publication is given in "Accredited Calibration Facilities".

2.3.9.3
International Activities

NKO/STERIN/STERLAB is a member of the European organisation EAL (European cooperation for Accreditation of Laboratories). EAL was formed on 31 May 1994 by a merger of the Western European Calibration Cooperation (WECC) and Western European Laboratory Accreditation Cooperation (WELAC). NKO/STERIN/STERLAB has the Chairman and Secretariat of the organisation.

This context is aimed at giving an international reputation to the national accreditation institutions. Through multilateral agreements, it aims at European acceptance of laboratory results. Within the EAL there are 2 multilateral agreements. One for testing and one for calibration.

The MLA for Calibration Laboratories has been signed by 11 countries: Denmark, Finland, France, Germany, Ireland, Italy, the Netherlands, Norway, Sweden, Switzerland, United Kingdom.

The MLA for Testing Laboratories has been signed by 12 countries: Denmark, Finland, France, Germany, Ireland, Italy, Norway, Spain, Sweden, Switzerland, The Netherlands, United Kingdom.

Since 21 October 1994 NKO/STERIN/STERLAB has also been Chairman and Secretariat of ILAC (International Laboratory Accreditation Conference).

Apart from these international activities NKO/STERIN/STERLAB considers it very important that apart from its own interests, the interests of laboratories is looked after when new international criteria are developed. This is illustrated by the fact that NKO/STERIN/STERLAB chairs both ISO CASCO WG10, responsible for the revision of ISO/IEC Guide 25, and CEN/CENELEC/ TC1/WG3, responsible for EN 45001-3.

Address:
Secretariat STERLAB
PO Box 29152
NL-3001 GD Rotterdam
The Netherlands
Telephone +31104136011
Telefax   +31104133557

2.3.10
The Polish Accreditation System

Z. Dobkowski and B. Berdowski

In Poland, at present, only the governmental accreditation system has been founded. Since 1st January 1994, the Polish Center of Testing and Certification (PCBC) has been in existence following the decision of the Polish parliament (Sejm) of 3 April 1993 on testing and certification. The PCBC is continuing the activity of former Central Bureau for Quality of Products (CBJW = Centralne Biuro Jakości Wyrobów) that in 1990, started the implementation of an international quality system according to the ISO and EN standards. These standards were recently implemented as the Polish Standards (PN = Polska Norma). Thus the CBJW was the first Polish governmental accreditation body. In fact, the Polish system for qualification of products' quality was founded in 1959.

The main areas of activity of PCBC are presented below:

1. accreditation of certification bodies and testing laboratories according to requirements of PN-EN 45000 standards;
2. certification of quality assurance systems according to the ISO serie 9000 standards, identical with the PN-EN 29000 standards;
3. certification of auditors; qualified canditates and auditors of PCBC are registered;

4. certification of product quality (including chemicals, petrochemicals, fertilizers, food products, rubbers and plastics etc.), as well as coordination and supervision of activities of certification bodies;
5. training of quality specialists on quality concepts, on accreditation of testing laboratories and certification bodies, on certification of products and on quality assurance systems.

In the PCBC there are three main divisions:

1. Bureau for Accreditation where the following sections are located:
   - accreditation of testing laboratories,
   - accreditation of certification bodies,
   - training and certification of auditors,
   - quality systems;
2. Bureau for Certification, where among others there are some chemical sections located, namely
   - for chemical and food industries (cosmetics, household chemistry, some rubbers and plastics, various food products),
   - for building materials and petroleum products (branch in Gdańsk),
   - for fertilizers (branch in Piła);
3. Bureau for Testing, including chemical testing laboratories of PCBC, that is the chemical laboratory in Warsaw, the laboratory for building materials and petroleum products in Gdańsk, and the laboratory for fertilizers in Piła.

Therefore, the PCBC is the accreditation and certification body for Polish analytical chemical laboratories.

The Polish system for qualification of products' quality was founded in 1959 and the present testing and certification system is continuing the former quality system, taking the ISO and EN standards, guides and requirements into account. The system of product certification is shown in Fig. 7.

There are 17 accredited bodies for certification of products, at least 6 of them are involved in testing of chemical products.

The PCBC is cooperating with several international organizations, such as IEC, ISO TC 176 "Quality Management and Quality Assurance", ISO/CASCO, CEN/CENELEC, EOQ, EOTC and EUROLAB.

Two Polish organizations of institutions are associated with PCBC: (1) POLLAB, the organization of Polish testing laboratories, and (2) "Polskie Forum ISO 9000", the organization of manufacturers. The chemical sections are active in both organizations.

Moreover, since 1992 the Polish section of EURACHEM-POLAND, is also active as the non-governmental organization of chemical laboratories interested in implementation of quality assurance systems, accreditation and certification and testing of chemical materials and products. The EURACHEM-PL is led by the Industrial Chemistry Research Institute in Warsaw. At present, the most important task of EURACHEM-PL is to disseminate information

   - on the quality assurance system,
   - on international documents for accreditation of chemical laboratories,

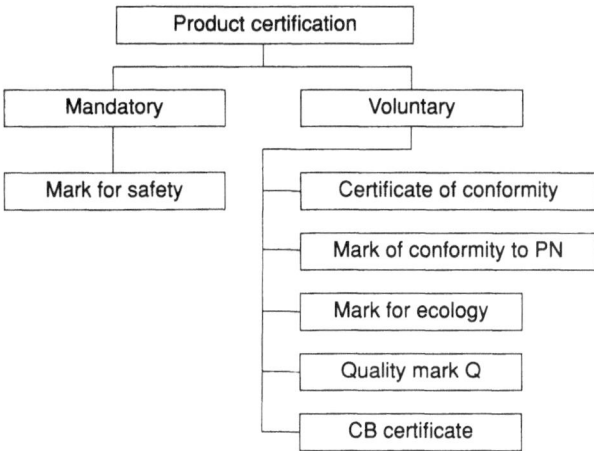

Fig. 7. The system of product certification

Explanations to Fig. 7:

Mark for safety – a mark indicating that the safety requirement for a product are fulfilled, i. e. the product is safe to life, health, property and the environment.

Certificate of conformity – it is awarded to denote conformity to requirements of standards, directives and regulations specified in the certificate.

Mark of conformity to PN – a mark indicating compliance with requirements of specified PN standards.

Mark of ecology – it is given to denote that the product does not produce negative effects on the environment during manufacturing, transportation, installation and use.

Quality mark Q – it means that the product exhibits reliability, use, ergonomic, health and organoleptic properties higher than average, and material and energy consumption lower than average.

CB certificate – it is awarded to denote conformity to requirements of IEC Publications.

– on European proficiency testing schemes,
– on problems of chemical measurements, including uncertainty of chemical measurements,

as well as an information on meetings, conferences and workshops sponsored by EURACHEM. Thus, close links of Polish chemical laboratories with the organization of international quality system are supported.

Addresses:
Polskie Centrum Badań i Certyficacji (PCC), ul, Kłobucka 23A,
02-699 Warszawa, Poland,
Telephone/Telefax (48-22) 47 25 01
Dr. Janusz B. Berdowski, Director of PCBC,
Telephone (48-22) 470742,
Telefax   (48-22) 471222

EURACHEM-PL, Industrial Chemistry Research Institute,
Rydygiera 8
01-793 Warszawa, Poland.
Prof. Dr Zbigniew Dobkowski,
Telephone (48-2) 6338298
Telefax   (48-2) 6338295

2.3.11
The Russian System for Analytical Laboratory Accreditation

Yu. A. Karpov[1], I.V. Boldyrev[2], G.I. Ramendik[3] and G.I. Freedman[4]

Chemical analysis is a specific measurement procedure that has a number of features different from other kinds of measurements. Firstly, analysis includes identification of the substance components, i. e., qualitative analysis along with a quantitative one. Secondly, the final stage of analysis is preceded by the sampling and sample preparation procedures. Thirdly, the analysis itself is usually accompanied by the chemical transformation of a sample to obtain an optimal analytical form. As a rule, uncertainties of the analysis "non-measurement" stages can significantly exceed the final determination's uncertainty. Finally, a special terminology for the chemical analysis has been developed as well as a special professional literature, special training system, etc.

All this has made it necessary to develop an advanced accreditation system for the analytical laboratories that would differ from other accreditation systems for testing laboratories and take into account the features of an analytical procedure mentioned above.

In 1980, a compulsory attestation for laboratories was introduced in the former USSR. More than 1000 laboratories has been attested over 5 – 6 years. When the USSR ceased to exist, the attestation was substituted by the accreditation of analytical laboratories (centres).

This accreditation system was developed in Russia by the Ural's Research Institute for Metrology in collaboration with the Association of Analytical Centres named "Analytica". The association includes over 100 analytical laboratories of the former Soviet Union, primarily the Russian ones. The system has been approved by the Russian State Committee on Standards. The system is based on the international (ISO) documents and European norm of the EN 45000 series.

---

[1] State Research Institute for Rare Metals, Moscow, Russia.
[2] Association "Analytica", Moscow, Russia.
[3] Institute of General and Inorganic Chemistry, Russian Academy of Sciences, Moscow, Russia.
[4] All – Russian Institute of Light Alloys, Moscow, Russia.

The new system began operating in October 1992. By 1 January 1995, 145 analytical laboratories had been accredited, among them laboratories conducting chemical analysis in the fields of:

metallurgy	18
agro-chemistry	60
petro-chemistry	6
environmental control	28
medicine	8

Access to the Accreditation System is free to any analytical laboratory.

A detailed experimental test for the competence of a laboratory is the main feature of the Russian Accreditation System. A special testing procedure is developed for each laboratory. Professional analysts, experienced in the field announced by a laboratory, form at least half of the Commission on Laboratories Accreditation.

2.3.11.1
Accreditation Criteria

1. Availability of conditions guaranteeing technical competence of an analytical laboratory in the accreditation field:
   - Availability of either necessary equipment or free access to such equipment;
   - Availability of reference materials, chemical reagents and materials;
   - Availability of the appropriate laboratory rooms;
   - Availability of the properly approved documentation (procedures, standards, instructions, etc.) necessary for the quantitative chemical analysis;
   - Availability of qualified staff;
   - Availability of the Quality Assurance system for the quantitative chemical analysis based on the principles, norms, regulations and procedures of the State measurement system and documented as a "Quality Manual".
2. Positive result of an experimental test for the quality of the analysis.

Independence (from the potential customers) of a laboratory can be recognised at the same time as the technical competence.

Conditions needed to recognise independence:

- Indicial status of a laboratory;
- No commercial interest of the laboratory in the results of the analysis;
- Impossibility of exerting pressure on the laboratory staff during the quantitative chemical analysis.

2.3.11.2
Accreditation System Structure

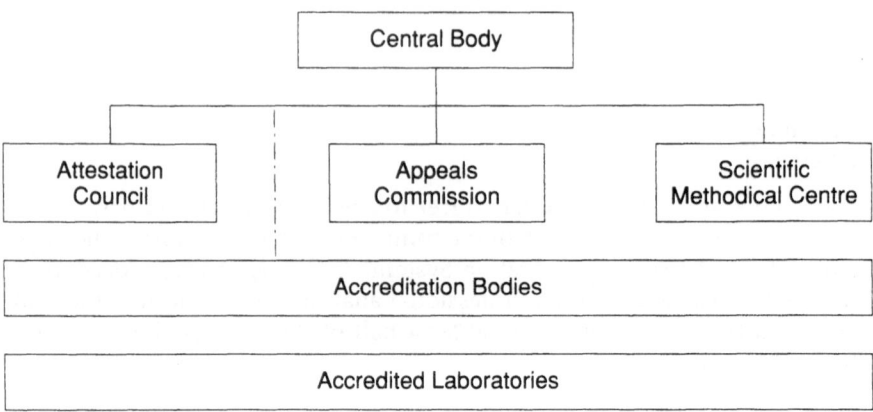

The Central Body is the organising and co-ordinating centre of the System. Its responsibilities include:

- Management of the operation of the Attestation Council and Appeals Commission;
- Responsibility for the maintenance of the System Register;
- Co-operation with the State management bodies on laboratory accreditation;
- Co-operation with the other Systems of Laboratory Accreditation.

The Attestation Council ensures that qualifications of the experts and auditors is judged impartially and manages their official appointments.

The Appeals Commission scrutinises appeals lodged against the decisions of Accreditation bodies.

Accreditation bodies ensure implementation of the whole accreditation procedures and take responsibility for the correctness of the judgement in terms of the analytical laboratory's conformity with the accreditation criteria. They also organise inspections of the operation of the accredited laboratories.

2.3.11.3
Accreditation Procedure

1. Handing in an application of the Accreditation Body.
2. Examination of an application. Making a conclusion.
3. Making up a Commission to test a laboratory.
4. The work of the Commission in the laboratory
5. Making a decision about accreditation.
6. Preparing and issuing an Accreditation Certificate.
7. Inspection control.

A laboratory is usually accredited for a period of 3 to 5 years. Details of the accredited laboratories are entered in the official Register.

Addresses of the Participants of the Accreditation System:

Yu. A. Bogomolov
Head of Committee of Russian Federation for
Standardization, Metrology and Certification
(GOSSTANDARD)
9, Leninski prospect
117049 Moscow

Prof. G. V. Ostroumov
Chairman of the Attestation Council
31, Staromonetniy per,
109017 Moscow

Prof. Yu. A, Karpov
President of the Association "Analytica"
9, Leninsky prospect
117049 Moscow

Dr. I. L. Dobrovolskiy
Director of Ural's Scientific Research
Institute of Metrology
4, Krasnoarmeyskaya ul.
620219 Ekaterinburg
Russian Federation

2.3.12
Accreditation of Laboratories in Sweden

Björn Lundgren

A control board for the appointment and the supervision of national testing laboratories was established in Sweden in 1973 giving certain national laboratories the authority of testing and inspection of specific objects of national interest.

2.3.12.1
The Accreditation Body

After a new governmental study in 1988 the Swedish Board for Technical Accreditation, SWEDAC, was established as a responsible agency according to the Swedish law (1989:164) "about control through technical evaluation and measurements". The accreditation process of national laboratories according to the previous regulation started 1989. The accreditation of other types of laboratory, especially for environmental analysis of waste water started very early and has since progressed in accordance with guidelines discussed and accepted in mutual cooperation between western European countries (see below). In connection with the negotiations between the EC and EFTA countries a system for mutual acceptance of test results had to be agreed upon. The Swedish governement announced Swedac as being the responsible authority for such a system in Sweden. In accordance with this decision, SWEDAC was also announced as having the whole responsibility for the accreditation. SWEDAC, in this way, became the central agency for technical control and measurement, the national agency for accreditation of laboratories, certification bodies and inspection bodies and carries the administrative responsibility for legal metrology according to EC and EFTA agreements. In its role as described, it became evident that SWEDAC is the national body for evaluation of laboratories for the notification to the EU-commission. The act of notification was made by the Swedish government.

SWEDAC has its main office in Borås and one office in Stockholm and employs presently (end of 1994) around 45 people. Lars Ettarp is the General director of SWEDAC. SWEDAC takes an active part in the international cooperation and development of guidelines, standards and regulations for accreditation and certification, particularly within the European organisations European Accreditation for Laboratories, EAL and European Accreditation for Certification, EAC. The cooperation of the nationally recognised accreditation bodies for testing started formally in 1989 and is based on a memorandum of understanding (MoU). An on-going program of cooperation was set up which aimed at establishing mutual confidence between bodies, so that agreements can be entered that recognised the technical equivalence of the operation of their accreditation systems for testing laboratories. SWEDAC have signed such agreements with Denmark, Finland, France, Ireland, Italy, The Netherlands, Norway, Switzerland, Spain and the United Kingdom. The agreement is published in STAFS 1994:11. A similar agreement has been signed with New Zealand (STAFS 1994:23).

2.3.12.2
The Accreditation Process

After application to SWEDAC for accreditation the documentation of quality coordinated actions and the performance of the laboratory is evaluated according to the requirements established, which are presented in series of documents announced as STAFS. The regulation for the accreditation of laboratories is announced in STAFS 1994:1 published 1994-02-04. These requirements are based on international harmonized standards, such as the EN 45 000-series and ISO/IEC Guide 25 (1990).

The evaluation is performed at the laboratory using one or several technical assessors as needed for the expertise and one accreditation assessor from SWEDAC. The accreditation assessor checks the status of the organisation, the leadership and responsibilities in the laboratory and audits the quality system and its applicability in various aspects according to the standard and from a customer's point of view. The technical assessor is an expert in the field and concentrates his review on the instrumentation, their handling and the methods and their applicability and the documentation of the validation procedures. The visit to the laboratory results in an Assessor's Report of the visit which normally presents an overview of the quality coordinated actions within the laboratory and in addition presents all the deviations from the requirements in the standard. After the fullfilment of the requirements, a certificate of accreditation is issued. The accredited laboratory is given an accreditation number and may use the logo of SWEDAC in their reports on accredited tests and analyses. When the logo is used for information purposes the type of accreditation and the standard applicable should be given below the logo. The requirements for the use of SWEDAC logo are published in a separate document.

An inspection of the laboratory performance is undertaken each year on site. This visit may, depending on the size and type of the laboratory, be performed by a technical assessor or an assessor employed by Swedac or both. The emphasis on this yearly inspection depends somewhat on the expertise of the appointed assessor. A complete reevaluation of the laboratory is performed every four years.

2.3.12.3
Types of Laboratory Accreditated

Presently the accreditated chemical laboratories have been divided into different types of laboratory such as for calibration, environmental control according to requirements of Swedish Environmental Protection Agency, Drinking water according to the Swedish Environmental Protection Agency, Drinking water according to the Swedish Agency for Food and Nutrition, General Chemical analysis, Clinical chemistry and General Microbiology. For other technical areas similar subdivisions have been made. The total number of accreditated laboratories of all types is close to 400. One laboratory may be represented in more than one group in the presentation below. This is in particular true for consulting laboratories.

2.3.12.4
Accreditated Laboratories in Different Areas (December 1994)

Calibration	75 laboratories
Environmental analysis	200 laboratories
Drinking water	60 laboratories
Outdoor air	20 laboratories
General Chemical analysis	20 laboratories
Clinical Chemistry	20 laboratories
General Microbiology	40 laboratories

All accreditated laboratories are kept on file at SWEDAC and periodically published by SWEDAC in a document presenting the laboratories, area of accreditation, technical area, property and object concerned.

Further information may be obtained from the SWEDAC offices below

	SWEDAC Head Office	SWEDAC office in Stockholm
Visiting address	Österlånggatan 5	Slussplan 9
Postal address	PO Box 878	PO Box 2231
	S-501 15 Borås	s-103 15 Stockholm
	Sweden	Sweden
Telephone	+46-(0)33-17 77 00	+46-(0)8-402-0070
Telefax	+46-(0)33-10 13 92	+46-(0)8-791-89-29

2.3.13
The United Kingdom Accreditation System

D. Galsworthy

2.3.13.1
History

During the 1970s the pressure for accreditation of test laboratories grew. This was in response to the demands of legislators, regulators and laboratory customers for an independent but competent third party who could judge laboratories against a set of agreed criteria acceptable to them as test customers and on whose judgement they could rely. The setting up of the National Testing Laboratory Accreditation Scheme (NATLAS) at the National Physical Laboratory ensued in 1980.

Around this time the national awareness in the principles of quality assurance was gathering momentum and there were signs of a marked developing in the supply of certification of quality systems in manufacturing enterprises and of the certification of products to BS5750 (ISO 9000). The British Government White Paper 'Standards, Quality and International Competitiveness' of 1982, set out the Government's approach to quality and proposed the setting up of the National Accreditation Council for Certification Bodies (NACCB). It laid down testing in NATLAS accredited laboratories as a foundation stone in the approach taken to product certification.

2.3.13.2
The Objectives of NAMAS

The two limbs of the UK laboratory accreditation system, the British Calibration Service and NATLAS were drawn together in 1985 to become the National Measurement Accreditation Service (NAMAS). The objectives then set out for the service which still pertain are:

- to establish the widespread recognition of the competence of accredited laboratories
- to improve the authority and standard of testing
- to eliminate multiple assessment of calibration and testing laboratories
- to develop mutual recognition agreements
- to provide publicity for accredited laboratories and a service to their users through the publication of a NAMAS directory of Accredited Laboratories.

2.3.13.3
The Formation of the United Kingdom Accreditation Service

On 1 August 1995, the United Kingdom Accreditation Service was formed by bringing together the work of NAMAS and the National Accreditation Council for Certification Bodies (NACCB, a former Council of the British Standard

Institute). It provides a unified national accreditation service for bodies undertaking certification of products, personnel or systems – including environmental management and audit systems – as well as laboratories performing tests and calibrations.

The body operates as an non-profit distributing "Company Limited by Guarantee" whose members represent the entire spectrum of national interests in accreditation.

In recognition of UKAS role as the UK's accreditation authority for certification bodies and testing and calibration laboratories, the UK Government has granted UKAS permission to continue to use the existing accreditation marks – the "tick and crown" and NAMAS logo.

Accreditation is awarded by UKAS to third parties independent certification bodies and testing/calibration laboratories as recognition that they meet internationally agreed criteria covering integrity, technical competence and validity of methods. This is set out in a Memorandum of Understanding between the Secretary of State for Trade and Industry and UKAS an it records their joint commitment to maintaining and developing a strong and unified national accreditation service in the UK as a means of promoting quality and improving the competitiveness of UK industry.

Present NAMAS accreditations remain valid and may continue to use the NAMAS (National Accreditation of Measurement and Sampling) logo. New accreditation assessment and surveillance visits will be made by UKAS, who will grant accreditation. UKAS will also continue to play a major role in EAL (European cooperation on Accredited Laboratories) and other European and World organisations.

UKAS operates in line with the international standards laid down for laboratory accreditation bodies (EN 45003/ISO Guide 58) and conducts assessments in line with EN 45002. At present, the laboratory accreditation wing of UKAS has a staff of 64 of whom 43 are technical case officers. However, the principal and specialist technical input to assessment is drawn from a pool of 300 national experts in various fields all of whom have undergone formal training in quality system assessment and in assessment to the accreditation standard.

UKAS assesses laboratories against the NAMAS Accreditation Standard M10, "General criteria of competence for calibration and testing laboratories" which is consistent with the equivalent international criteria documents EN 45001 and ISO Guide 25. The standard requires interpretation as it is applied to an increasing range of diverse fields and sector specific guides have been produced in consultation with the laboratory and use community.

UKAS fully recovers the cost of its accreditation programme, through fees charged to laboratories. Fees are based on the amount of assessor effort need to take a laboratory from application to accreditation and thereafter to conduct annual surveillance and periodic reassessment. Fees are independent of overall turnover of the enterprise and geographic location in the UK.

At the beginning of September 1995, the laboratory accreditation side of UKAS has currently 1818 accreditations, consisting of 1281 testing and 537 cali-

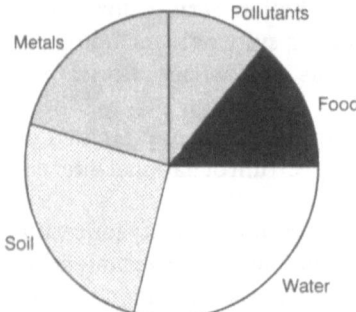

Total number of NAMAS UK chemical
laboratory accreditations – 420 laboratories

bration accreditations. New applications are being received at the rate of approximately 200 per year and the major growth areas remain chemical analysis, microbiology and construction site work activities.

2.3.13.4
Legislative Support for Accreditation in the United Kingdom

Although the UK Government has stated its support for the use of accredited testing and has said that it should set an example by using accreditation, the UK does not, unlike some countries, have legislation which requires all testing in support of regulation to be conducted in accredited laboratories. UKAS aims to achieve this through persuasion. This line has nevertheless been successful. The Department of Transport requires all materials used in motorway construction projects to be tested in accredited laboratories and requires all site investigation activities to be accredited. The Drinking Water Inspectorate of the Department of the Environment accepts UKAS accreditation of drinking water analytical laboratories as part of its own assessment of drinking water supplies. The Health and Safety Executive has strengthened significantly its advice regarding the use of UKAS laboratories in the codes of practice associated with Asbestos Regulations and the Inspectorate of Pollution are stipulating the use of UKAS accredited laboratories for stack emission compliance monitoring activities. The UK implementation of EC Directives is also making considerable use of the accredited laboratory infrastructure, particularly in relation to the testing of toys, personal protective equipment, food and child resistant containers.

2.3.13.5
Relationship between Laboratory Accreditation and Certification to ISO 9000

A number of organisations offering testing and calibration also provide other products and services such as research and development or consultancy. ISO Guide 25 meets the requirement of the ISO 9000 series of standards when acting as suppliers producing test or calibration results. However, there is often a need for certification of the quality system relating to the other areas of laboratory activity. There is no reason why this cannot be provided and there should be no need for any duplication of effort or assessment by UKAS and the chosen certification body. UKAS is already gaining experience with at least one certification body of joint NAMAS/ISO 9000 assessment of laboratories and looks forward to refining this approach in the interest of the customer.

2.3.13.6
Co-operation Between UKAS Laboratory Accreditation and the UK Good Laboratory Practice (GLP) Monitoring Unit

The UK GLP Monitoring Unit is administered by the Department of Health and is responsible for dealing with all enquiries regarding GLP from the UK and overseas regulatory authorities and from the GLP monitoring units in other countries.

The GLP Monitoring Unit was established in order to carry out inspections and study audits on chemical substances where the results of the studies are to be submitted to regulatory authorities. These inspections are carried out to establish if laboratories have satisfactorily implemented the internationally agreed principles of Good Laboratory Practice. As a result of the discussions between NAMAS and the GLP Monitoring Unit, it was recognised that close similarities existed in the requirements which laboratories must meet before being included in either scheme. An agreement between the two schemes has been formalised to cover chemical and physical testing of products for data submission to regulatory authorities. UKAS and the GLP Monitoring Unit are now cooperating in assessments in order to maximise efficiency and minimise costs for laboratories wishing to hold both GLP compliance and UKAS accreditation status.

The main differences in the requirements of UKAS and the GLP Monitoring Unit originate from the difference in emphasis placed on the various aspects of a laboratory's organisation and have been explained in the information sheet detailing the agreement between the two organisations.

2.3.13.7
Areas for Development

UKAS and its approach to accreditation is always undergoing development to meet the changing needs of its laboratories and their custumers. Apart from the development of guidance documentation to accommodate new fields of accreditation, UKAS, together with its European partners, is considering its approach to scopes and schedules, measurement uncertainty, its interface with the certification bodies, proficiency testing and accreditation criteria. Significant progress has been made recently in the UK on a generic aproach to the definition of the accredited scope in the field of analytical chemistry.

2.3.13.8
Current Concerns

As UKAS grows in size a particular concern is to ensure that all technical assessors apply the accreditation standard in a consistent way. UKAS has embarked upon a series of assessor updating seminars to ensure uniform application and full awareness of recent developments.

2.3.13.9
Case study: NAMAS Accreditation and the "Additional Measures Directive"

On the 29 October 1993, the European Council adopted a Directive on the Additional Measures concerning the Official Control of Foodstuffs (93/99/ EEC). The aim of the Directive is to ensure food enforcement results obtained by Inspection bodies in the different Member States are produced by competent bodies. In order to safeguard the interests of both human health and of legal security and inspire consumer confidence, mandatory requirements have been specified that the laboratories used to produce the results are notified and work to recognised standards of competence.

In addition to being authorised to carry out formal analysis and examinations for the purposes of enforcement of the UK Food Safety Act 1990, notified laboratories will have to comply with the general criteria for the operation of testing laboratories as laid down in European Standard EN 45001. The bodies responsible for assessing the laboratories comply with the critria laid down in the European Standards EN 45002 and EN 45003. In the United Kingdom, UKAS Executive meets the criteria of EN 45002 and EN 45003 and laboratories accredited by UKAS meet the requirements of EN 45001.

The Directive also indicates that food enforcement laboratories must participate in prescribed analytical proficiency schemes. One such scheme is the Food Analysis Performance Assessment Scheme (FAPAS) organised by the Ministry of Agriculture, Fisheries and Food in the field of the chemical analysis of foods.

For the purposes of the Directive, Member states will shortly present the Commission with details of the notified laboratories. In the UK these will include NAMAS accredited Public Analyst and Public Health Laboratory Service laboratories.

Address:
United Kingdom Accreditation Service
Queens Road
Teddington
Middlesex
United Kingdom TW 11 OLW
Telephone 44-181-943 6311
Telefax   44-181-943 7134

Chapter 3

Quality Assurance in Analytical Chemistry

Karl Heinz Koch

3.1
On Quality Assurance

The triad quality, testing, and progress is a decisive factor when facing future chalenges of the market (Fig. 1). Therefore, a standing objective is to work on "quality culture", i.e. to optimize the combination and coordination of all quality-relevant individual corporate functions. Quality assurance measues are an essential component of future-directed industrial trading. The three parameters: price, quality, and flexibility, are having an increasing influence on the market (Fig. 2). Hence, quality assurance of marketable products becomes part of the quality policy of industrial companies.

In this context, quality, defined as the "the features and characteristics of a product or an action in their entirety to fulfill predetermined requirements" (DIN (German Standard) 55350, part 11), has to be quantifiable. This quantification is sustained by quality inspection in the frame work of quality assurance. Chemical analysis contributes to these quality ensuring measures in an important, in many cases overriding and decisive way [1]. In order to meet the requirements of these tasks of product-oriented testing in the crucial field of quality policy and economy, the documented and verifiable integration of quality assuring measures into the whole analytical process is needed [2].

Further reasons for these activities are a changing sense of justice and related changes in the legal situation. Shifting the burden of proof of the producer's product liability [3] (German law on product liability of 01.01.1990) and the changed public awareness of the environment have brought questions on "quality assurance" related to the production and use of materials or machinery into the open.

Quality assurance concepts cover different corporate sectors and should start of the earliest possible stage because correcting mistakes or deviations

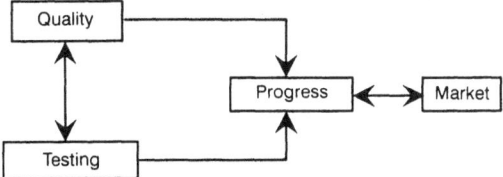

Fig. 1. Future-directed factors influencing the market

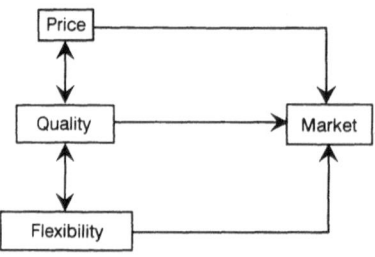

Fig. 2. Market determining parameters

from this intended production target at the final inspection stage satisfies neither producer nor customer. In international competition, the quality of products and services is becoming increasingly significant. Thus, quality, is becoming a competitive instrument for companies (e.g. the slogan: "We create quality!"). Investment in quality assurance measures is considerable because quality management systems have to be set up in a product- and company specific way. Only in this way it is "assured" that "quality" is not an accidental result.

The fundamental terms of quality assurance have to be defined unambiguously in order to avoid linguistic mistakes and misunderstandings; they are themes in the already mentiond standard DIN 55 350, part 11 and of DIN ISO 8402 (draft). According to these, the following definitions (in part, only their gist is cited) apply.

Quality management:	All activities of the general management that determine the quality policy, the aims and responsibilities as well as their realization by means of strategic planning, quality assurance and quality improvement in the framework of the quality management system.
Quality policy:	The overall quality intentions and direction of an organization as regards quality, as formally expressed by top management.
Quality planning:	Selection, classification and weighting of quality features as well as stepwise concretization of all individual quality requirements with regard to specifications for achieving them.
Quality control:	Preventive, surveillance and corrective measures for the realization of a unit (this may be: the result of an activity or process) with the aim of fulfilling quality requirements.
Quality assurance:	All those planned and systematic actions, realized within the quality management system, necessary to provide adequate confidence in the fact that a unit will satisfy the requirements for quality.
Quality management system:	The organizational structure, responsibilities, proceedings, processes and required resources for the implementation of quality management.

Quality inspection: Determines how far a unit (e.g. products or services) meets the quality requirements. Or, in other terms: System of testing methods, intended to maintain a predetermined quality.

3.2
Quality Policy and Quality Management

3.2.1
Corporate Quality Policy and Quality Strategy

Today, more than ever, product quality presents a challenge to the top management of a company, because it determines productivity and profit not only at present but also in the future. In the framework of its quality policy, management has to ensure that *all* quality relevant activities are synchronized in order to achieve the optimum total result. Thus, quality policy becomes a fundamental component of corporate policy, the realization of which is a direct responsibility of management. This aspect is embodied in the related standards. The above-quoted standard DIN 55 350 (part 11) is completed by the statement of DIN ISO 9004 which says (among other things):

"Management should take all the necessary measures to ensure that its corporate quality policy is understood, implemented and maintained."

In order to realize the objectives of quality policy, a quality strategy is needed which is expressed by certain basic fundamentals of the company. This includes formulations enabling each empoyee to understand the significance of quality-aware measures. In the competitive sphere of our industrial society, one of the most important future managerial tasks will be to initiate and maintain this awareness.

The fundamentals reflecting corporate quality strategy, of course, comprise statements on meeting deadlines and on the maintenance of customer services. Customer expectations will be the standard measure for corporate quality, where "quality" is the result of the cooperation of all the employees. Thus, quality is developing into a key factor for the existence of companies and for safeguarding their future. Personnel resistance to QA measures is most effectively avoided or removed by a manifest and clearly expressed promotion of the quality concept by management. Interdisciplinary communication and modern decision-making techniques are the instruments used to avoid undesirable developments and failures.

Quality policy is submitted to a constant process for improvement. This permanent process enforces constant concern with regard to quality and productivity questions, creativity and innovation, in order to meet the perpetually changing requirements of the market. Here, corporate long-term success depends on satisfied customers, satisfied staff and flexible organizational structures.

3.2.2
Quality Management and Quality Assurance

Quality management covers all the actions related to quality planning, quality control, and quality assurance. Formerly, industrial quality assurance was product-related, today, this still remains the main focus, but, by now, in general a process-related organization exists. Farsighted companies are switching over to system-related quality assurance, by which sources of error are clearly defined and eliminated interdepartmentally. The last mentioned way of proceeding presupposes a central quality organization with appropriate competence, which enjoys a coordinating function and which is in charge of internal auditing within the company.

Thus, product quality is "created" in all company operations; each operation has to take over responsibility for the quality element of its area. In pre-production fields like marketing, planning and development, focus is on avoiding mistakes. Purchasing policy resulting from this type of quality assurance concentrates on the supplier's quality capability. Certificates from a third part quality assured production of the relevant product may be extremely beneficial to business relations.

In summary, the following can be stated: quality originates from a perfect combination of the factors man, machinery, method and material (Fig. 3). Effective quality assurance has to be applied in all four sectors. Materials may be tested on delivery completely, item by item, or by random sampling. The

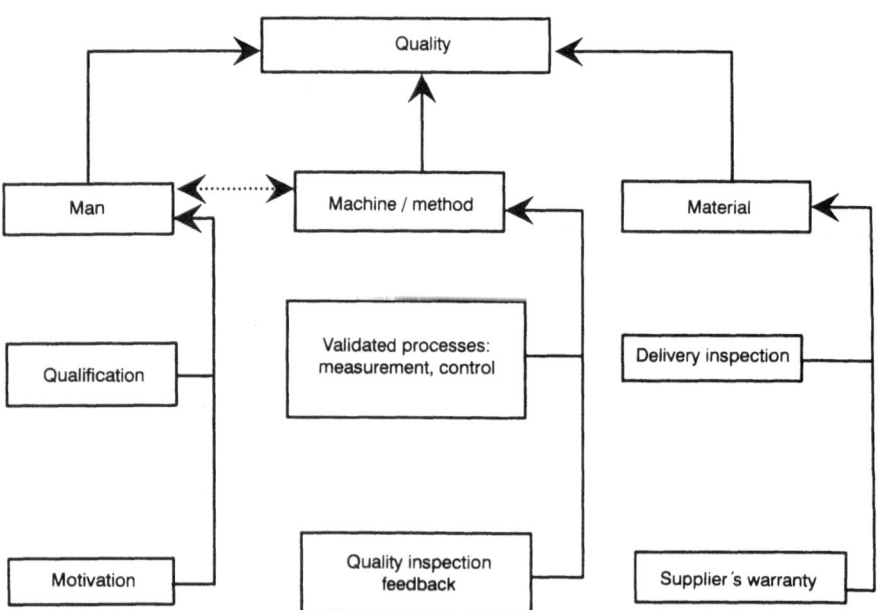

Fig. 3. Interrelationship of man, machinery, method, and material

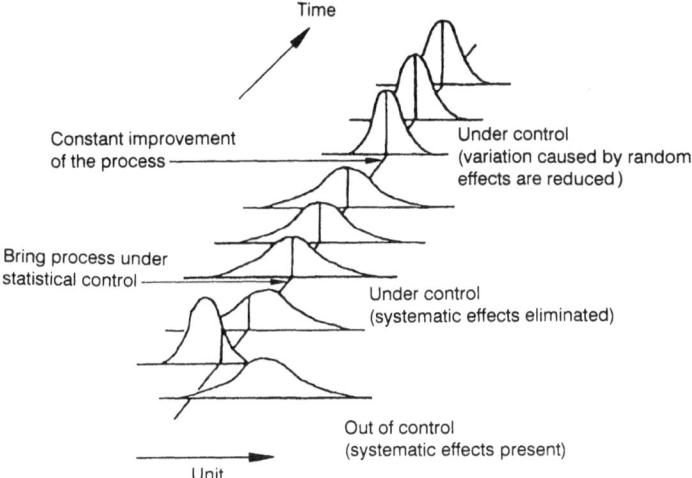

Fig. 4. Effect of process control

outlined new purchasing policy ensures quality; in regular terms, a serification of quality assurance at the supplier's is foreseen (auditing). Machinery and methods used in the production process are characterized by computer-aided measurement and control engineering, which guarantees production safety to a high extent. Nowadays, statistical process control facilitates the quality-dependent on-line control of processes (Fig. 4). In this context, the most important uncertainty factor is man. He operates, supervises and maintains machinery and installations, checks product quality and makes decisions. While doing so, of course, he may make mistakes. Therefore, all employees of a company have to gain the knowledge, necessary to fulfill their tasks in a quality-aware manner. Through more willingness to cooperate and a stronger identification with the task, this expenditure leads towards greater statisfaction of the individual, better economic viability of the company, and, thus, higher job security.

3.2.3
Total Quality Management (TQM)

Looking for appropriate management tools that match the above mentioned challenges, the term and method of Total Quality Management (TQM) was created, which may be conceived as quality aware comprehensive management [8]. The objectives of TQM are as follows [8]:

- secure profit through quality orientation
- increased competitiveness due to higher quality
- intentional increase in customer satisfaction
- involvement and motivation of staff members

- complete control over products, services and processes (zero defect)
- strict observance of promised delivery dates and quantities (just-in-time)
- reduction in losses caused by errors, rejects, corrective and additional work
- better utilization of human, technical, and organizational resources.

Implementing quality awareness at all levels generates social responsibility so that questions on environmental tolerance and safety are tackled with enthusiasm and solved. The practical implementation of TQM is characterized by:

- customer orientation,
- staff orientation,
- social orientation,
- creation of effective programs for the realization of TQM.

In this way, social orientation becomes a part of quality management. For extensive information on these subjects and possible methodologies, a consultation of the specialized literature is recommended [9–13].

3.2.4
Quality Costs

Apart from expenditures for raw material, resources, personnel etc., every manufacturer incurs expenses resulting from the achievement of the required quality: the quality costs. According to their causes, these costs may be divided into four groups (Fig. 5):

- costs involved in the prevention of faults and errors
- costs arising from testing
- costs caused by faults occuring within the company (internal costs),
- costs resulting from faults occuring after delivery to the customer (external costs).

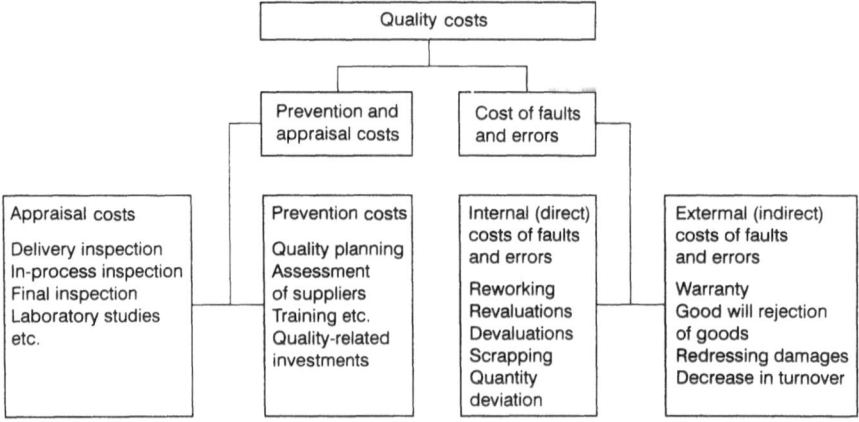

Fig. 5. Breakdown of quality costs [7]

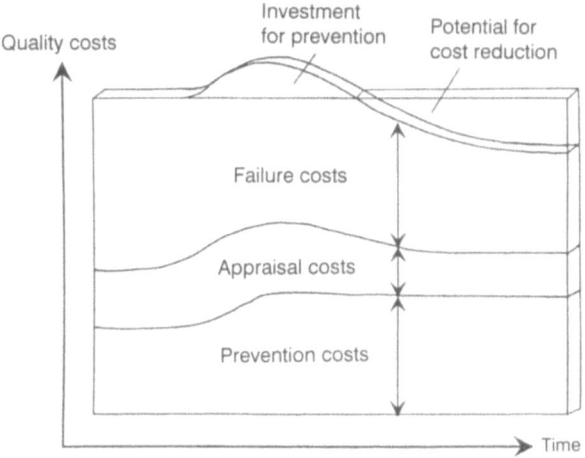

Fig. 6. Dependence of quality costs on the cost of faults

How high these quality costes are can be influenced not only by improved production- and control processes, more precise measuring methods, but also by a greater commitment, a stronger sense of responsibility and better professional training and information of the employees. A reduction in the total quality costs may be realized by reducing costs due to faults and errors (see Fig. 6).

3.3
Quality Planning, Quality Control, Quality Inspection

In the framework of quality planning, quality features which the product should comply with are defined and evaluated with regard to technical conditions and customer requirements. The quality inspection allows one to determine whether a product or service meets the quality requirement; it is subdivided into planning and carrying out of the study and data processing (Fig. 7). Quality control, finally, uses test data resulting from the quality inspection for controlling "quality production", and, if necessary, applies corrective measues to a running production process.

From this short description it follows that with overall quality assurance, the main consideration should be quality inspection because its individual activities cover all production steps encountered in a company or production unit.

Figure 8 shows, as an example, the course of a quality inspection process for steel production. At the same time, this example elucidates the special role chemical analysis plays in the framework of quality-assured production.

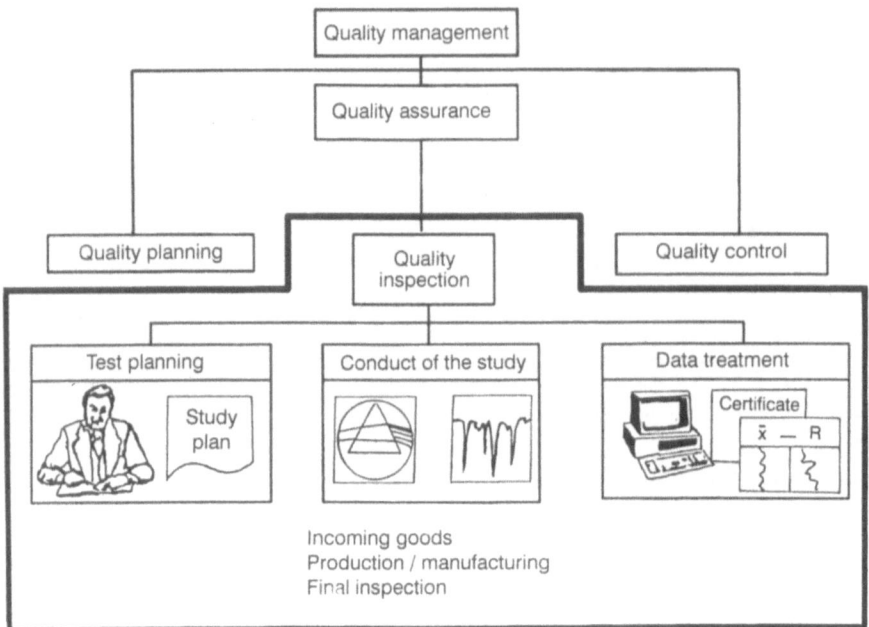

Fig. 7. Organization of quality assurance

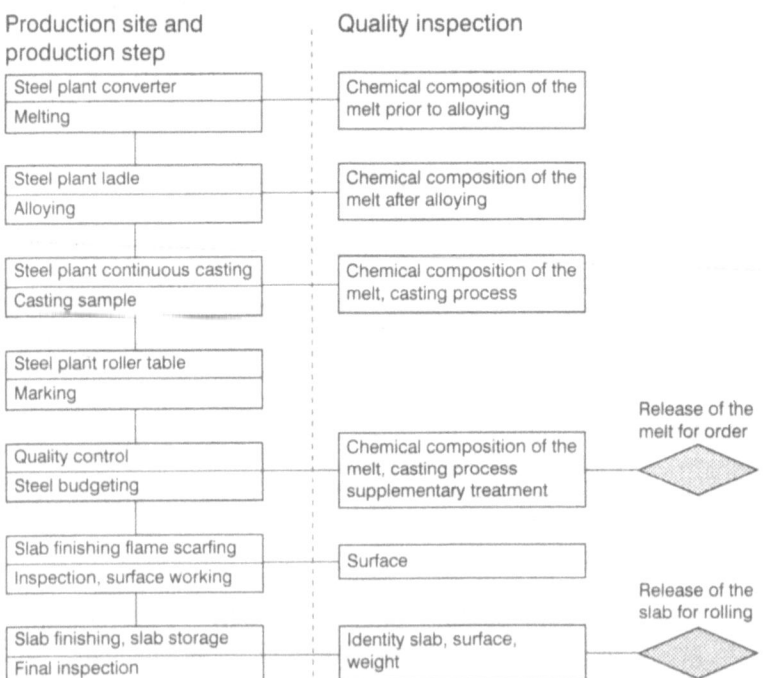

Fig. 8. Quality Inspection for steel production

3.4
Quality Assurance in Analytical Chemistry

3.4.1
The Significance of Quality Assurance For and In Chemical Analysis

In chemical analysis, as in many other fields, the use of validated methods of testing quality in order to ensure the accuracy of test results, has become a subject of increasing and fundamental significance.

Here, chemical analysis and quality assurance are combined. On the one hand, analysis is an integral part of quality assurance and provides data needed within the framework of quality assurance to give the desired results [15]. On the other hand, these analytical data must be validated by the relative integral QA measures [16].

For a few years, there have been several international guidelines and standards stipulating requirements for the competence and acceptance of testing laboratories. In this context, the following should be mentioned

- ISO Guide 25: "General requirements for the technical competence of testing laboratories", and,
- ISO Guide 38: "General requirements for the acceptance of testing laboratories".

The logical complement to these guidelines is ISO Guide 49: "Guidelines for the development of a Quality Manual for a testing laboratory". These give an outline for compiling a quality manual and it includes, as an essential part, requirements for the provisions to be made with regard to the application and monitoring of measuring and test installations (equipment and materials) as well as the recording of related data [17].

This process for setting up quality manuals for chemical analytical laboratories is in full progress and probably already completed in quite a number of cases.

The following European standards represent another basis for QA-related activities (see chapter 2 [18]):

- EN 45001   General criteria for the operation of testing laboratories
- EN 45002   General criteria for the assessment of testing laboratories
- EN 45003   General criteria for Laboratory Accreditation Bodies.

These standards complete the standard series ISO 9000 [5, 19] or EN 29000 which constitute the guidelines for the selection and application of standards for quality management, for the elements of a quality assurance system, and, for quality assurance control levels.

In the framework of product quality assurance, chemical examination is usually included in the product-oriented measures for quality inspection. As already mentioned, the results of chemical analytical investigations are part of the measurement results and assessment criteria which flow into the quality management system and contribute to a coordinated quality assurance; with regard to inividual phases of production processes, they may be even of *crucial* significance.

From this it is clear that test methods should be used which determine the chemical compositon with a precision related to the particular requirements. Furthermore, in some circumstances, delays will also have to be considered. Obviously, the most economic procedures will be applied if they fulfill the requirements.

Like all test methods, methods of chemical analysis are not free of errors. It is known that the occurring total error is comosed of a systematic and a random contribution (see chapter 5 [20]). The efforts of everyone involved with the production of a product should be aimed at minimizing systematic and random errors. Beside the necessary technical installations, this requires highly qualified staff and goal-directed personnel management; both of which will be discussed in the following.

3.4.2
Consequences for Quality Assurance in Analytical Laboratories

3.4.2.1
Compiling a Quality Manual

Results of chemical analysis constitute – as already mentioned – an important part of measurement results and assessment criteria which contribute to a company's product-oriented quality assurance.

Measures for quality assurance in the analytical laboratory, and the related instrument specifiation etc., are written down in task-oriented quality manuals (QM); the extent and content of which is defined by the above-mentioned standards and guidelines. The QM, serving as an internal company rule book is the basis of every system audit; special attention has to be drawn to the fact that all the information relevant to a system audit is to be take from it or refer it.

Drafting of a quality manual thus requires special knowledge of the fundamentals and requirements of quality assurance in order to meet these requirements. Seminars, offered by various bodies are intended to fill gaps in knowledge in the field of quality assurance and to provide help for drafting a quality manual.

A QM for analytical laboratories is sub-divided into sections; the following sub-division and sequence may be considered as an example:

Revisions
  0. Table of contents
  1. Quality statement
  2. Scope
  3. Basis of the quality management system
  4. Premises and installations
  5. Organization

6. Personnel qualifications
7. Procurement of measuring and testing equipment and testing agents
8. Measuring and testing equipment and testing agents
9. Checking measuring and testing equipment and testing agents
10. Test control
    - Sampling
    - Sample labelling
    - Sample transport
    - Sample preparation
    - Cross-reference to toher organizational units
11. Testing
12. Quality Assurance
    - Control analyses
    - Certified Reference Materials
    - Round robin analyses
    - Strategy for emergencies
    - Internal quality audits
13. Documentation
    - Test reports
    - Updating service
14. Pertinent documents, regulations and guidelines.

In the following, the terms used as titles of the individual sections will be explained or will be supplemented by an example:

The list of *revisions* should indicate the effective state of the QM showing for every individual section the currently valid version with its date of release.

Furthermore, a list of everyone possessing a numbered copy of the QM should be available at any time. A copy of the signed receipt of each person who possesses a QM is kept together with the mentioned list. The QM is confidential and may be neither copied in total nor in part without the express permission in writing from the department responsible for issuing it. The *table of contents* outlines the topics of the QM and is, like all other pages, is under control of the updating service.

The introductory implementing declaration defines the scope of the QM; it contains, furthermore, the commitment of the responsible company department to adhere to the described quality system.

A possible formulation of this commitment may read:

This QM describes in detail the quality management system (QMS) which is valid for the analytical laboratory within... (The department of The management)... commits itself to follow QS described herein. Staff of the analytical laboratory are instructed to carry out their tasks with due regard to the obligations laid dow in the QM.

This is followed by the name(s) of the person(s) or department(s) involved and the signature(s) of the person(s) responsible.

The *scope* of application within the company and the *basis* of the QM system are subsequently explained.

In the scope of application, it is usually specifed that the purpose of the quality system is to achieve and to maintain a high standard of quality for all testing activities. It is part of the quality policy of the analytical laboratory itself in the case of an independent institution, or of the company to which the analytical laboratory belongs.

The QM describes the elements of the quality management system (QM system) for chemical testing and their realization in the individual working areas. All the staff are obliged to observe the regulations set down in this manual for their field of work. The department head is responsible for conducting tests according to the QM. The QM contains fundamental statements on instructions for testing and documentation, and, further, the description of specific testing methods and procedures which serve to ensure the quality of testing.

Neither the work of an analytical laboratory should be subject to influences restricting technical judgement, nor should any outside person or organization be in a position to influence the investigation and results. Staff remuneration should depend neither on the number of effected tests nor on their results.

The standard series DIN ISO 9000 through 9004 [18], EN 29000 through 29004, or EN 45000 ff [18] constitute the basis for the QM system described by a QM. Furthermore, the principles of Good Laboratory Practice (GLP Principles) embodied e.g. in the German Chemikaliengesetz (Gesetz zum Schutz vor gefährlichen Stoffen – ChemG, German hazardous chemicals legislation of September 16th, 1980, BGBl I, p. 1718ff., amended on September, 15th 1986, BGBl I, p. 1505ff) are taken into consideration.

Further sections cover provisions on the used *premises and installations* and on the organization of the concerned department (flow-chart).

The provisions on *premises and installations* should contain the following general statements:

In order to guarantee that investigations can be conducted free from outside influences, the premises, for which plans of the building are available, conform to the laboratory guidelines issued by the main association of the Employment Accident Insurance Fund of the Chemical Industry, or the regulations governing workplaces. Suitable steps have been taken to ensure that influences from the surrounding do not impair the required quality of analytical work. For special work, fume-hoods are available allowing, in special cases, work, for example with acids or organic solvents, to be carried out. Lighting of the work places is in accordance to the regulations and is dazzle free. The storage of chemicals as well as labelling and storage of containers follows GLP rules or the respective regulations. Apart from the laboratory management safety personell ensure that laboratory safety regulations are respected. Signs warn people that unauthorized entry to the chemical testing rooms is forbidden.

The section on *personnel qualification* contains the valid quality criteria applicable to each organizational level. It is completed by documentation of job descriptions filed according to the field of activity, and, by certificates of further professional training and training courses (see Sect 3.4.2.1).

Finally, the QM contains information on the *documentation* of test reports and on change control. After consultation with the customer, test results are transmitted as a written report by post, or by means of telex, telefax, or by electronic data transfer. The individual company departments keep these test reports for a certain period in written or electronic form. The period of safekeeping depends on the type of tested material. Information on this topic it generally included in documentation kept in the laboratory department concerned.

Every QM, including all related documentation and instructions, is subject to revision. All documents are revised annually. The responsible person for this updating service is the manager responsible for the analytical laboratory. He can call for assistance from any of his staff who are directly concerned. After review or modification, he informs the central unit for the quality assurance of the company and all the people concerned who are his responsibility on the result of the review or he supplies them with the changed pages of the QM. These changes are documented and one copy of the old version is saved.

The list of revisions gives an account of the modifications which have been made. The relevant *documents, regulations, and guidelines* are listed at the end of the QM. These are ISO/IEC Guides [22, 23], ISO and EN standards [24] as well as, e.g. in Germany, VDI/VDE/DGQ guidelines [25] mentioned above or in the following reference list, and, if necessary, internal operating instructions.

The various analytical measures described in the QM will be discussed in chapter 5.

3.4.2.2
Personnel Qualifications and Equipment

The success of quality assurance measures, like the positive application of modern analytical techniques, requires qualified – and motivated – specialists. Apart from professional experience, the required personnel qualification (see Sect. 3.4.1) is obtained through job-oriented internal and external training; its extent is documented and may be verified at any time. High-tech testing methods applied by either inexperienced or for unqualified staff members could lead to serious problems for industry and society because of the "wrong" results obtained and the associated misinterpretation. Here, "tester" is the link in a chain which extends from materials development helping technical progress to quality-assured production of goods which meet market demands.

Laboratory equipment and personnel qualification rank equally. In order to carry out the laboratory's tasks, efforts are constantly made to make use of the most modern technology. This leads to highly specialized analytical equipment to meet the demands of the product, and to a high level of knowledge and experience of all the workers. The organization and the equipment of an analytical laboratory (see Sect. 3.4.2.1) should not only allow an optimum completion of as assignment, but – as already stated – should be in accordance with quality management demands and, if necessary with legal requirements (e.g. checking measuring and test equipment, performance of control analyses among others). For laboratories, this may mean either an extension of their field of activity or totally new tasks with for which new methods and techniques may have to be developed.

The progress made in the manufacture of scientific apparatus has two significant aspects for the validation of analytical results: on the one side, the application of instrumental methods greatly reduces the number of steps in analysis compared to chemical procedures; this leads to a reduction in random and systematic errors. On the other hand, the use of computer-aided equipment leads to more accurate results through the prevention of errors generated when reading instruments, when calculating and by mistakes in transcription.

As a result, great care has to be given to the *purchase of measuring and test equipment*. Prior to purchase of an analytical or laboratory device (measuring and test equipment), the suppliers' "quality capability" is assessed. Selection criteria are related to the specific measuring and test equipment. This may that the manufacturer performs extensive analytical investigations on the basis of samples provided by the company.

On delivery, all measuring and test equipment should undergo a thorough inspection. This inspection should include at least: a visual inspection of the items for any damage, a verification of the supplied items for conformity to the requirements specified in the purchase order, and, if appropriate, an examination of the supplied documentation. Measuring and test equipment that does not meat the requirements may not be put on the inventory.

Chemicals needed for testing are ordered at reputable manufacturers on the basis of their quality declarations. The reliability of these chemicals is constantly assessed by a determination of the analytical blank. All analytical and laboratory devices (measuring and test equipment) which are used in quality-assured analysis are recorded in an adequate manner in the QM. These records should include the following information:

- description of the measuring and test equipment
- manufacturer, type, serial number
- year of manufacture
- operating conditions
- instrumental modifications (type, operator, date)
- analytical applications areas
- maintenance documents and information on maintenance intervals.

Instrument files used as documentation in a QM, are generally classified according to the product and the application and contain instrument specifications, operating conditions, and analytical application areas. Maintenance instructions and maintenance certificates are part of the documentation on the checking of measuring and test equipment (see Sect. 3.5.1).

In addition the devices and systems currently used for "instrumental" analysis, standardized laboratory equipment is used, e.g., in the case of German Laboratories:

aerometers	DIN 12790 and 12791,
burettes	DIN 12700,
Erlenmeyer flasks	DIN 12380,
volumetric flasks	DIN 12664,

graduated cylinders	DIN 12680 and 12685,
filter papers	DIN 12448,
pipettes	DIN 12687, 12689 and 12691,
round- and flat-bottom flasks	DIN 12347,
thermometers	DIN 12775, 12778, 12781, 12784,
	12785, 12789, and,
crucibles	DIN 12904.

Analytical balances needed for chemical studies are purchased according to required precision criteria and are checked at regular intervals. In a documentation related to this section, all balances should be recorded according the following criteria:

- manufacturer, type, serial number
- year of manufacture
- current location
- timetable for checking and checking documentation

3.5
QA Measures in Analytical Practice

3.5.1
Checking Measuring and Test Equipment

Apart from records related to measuring and test equipment, a QM furnishes proof of their correct use. This set of problems includes the checking of measuring and test equipment, the recalibration of measuring and test equipment (calibration of a measuring and test device by means of recalibration samples), the determination of recalibration intervals and of criteria for the evaluation of analytical results in the framework of measuring and test equipment checking. Though the description of the calibration procedures (calibration) constitutes part of the operating instructions, the instructions for recalibration are a subject of separate documentation. Calibration of measuring and test equipment is regularly documented. When results are unsatisfactory, measures are taken to remove the equipment from use and have it repaired. Recalibration results are documented in form of control charts [26] (example see Fig. 9).

The balances used for QA are inspected at fixed intervals by means of an officially calibrated set of reference weights. Results are to be documented (see Table 1).

3.5.2
Test Control

Test control affects sampling, sample labelling, sample transport and sample preparation. It is influenced through its cross-links to other organisational units (ordering company divisions or external customers). Sampling is done

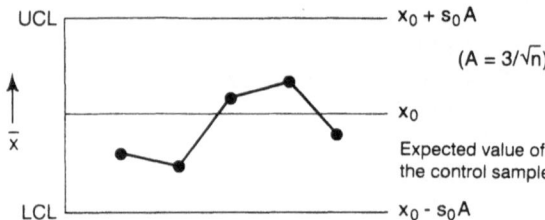

Corrective measure if:

1.) $\bar{x}$ lies outside the control limits
2.) 7 successive values of $\bar{x}$ are either located on one side of the center
line or show continiously increasing or decreasing values

Fig. 9. Control chart principal

Table 1. Example of the document used for recording the regular checking of analytical balances

Data Sheet Analytical Balance					
Type	XY				
Serial number	....				
Range	max. 110 g				
Sensitivity	....				
Used Reference Weights	nominal weight		admissible deviation		
Reference Weight 1	10.0000 g		± 0.0010 g		
Reference Weight 2	1.0000 g		± 0.0003 g		
Reference Weight 3	0.1000 g		± 0.0002 g		
Results of Control Weighings					
Date	Weight 1	Weight 2	Weight 3	Release	Signature
15.01.19					
15.02.19					
15.03.19					
15.04.19					
...					

either by the customer or by the analytical laboratory. In each case, highly specialised analytical expertise is required (refer to chapter 4 [27]). An undisturbed work-flow may only be guaranteed if the QM contains unambiguous provisions and procedures for sampling, sample labelling (identification) as well as transportation.

In general, the analytical laboratory always prepares samples; this is done in a substance-related as well as an analytical problem oriented way by following predetermined procedures described in the QM.

3.5.3
Testing (Test Instructions)

In the framework of QA within the analytical laboratory, unambiguous test instructions (description of the analytical procedure) should be established for all components to be determined in the materials under QA. These instructions include data on the equipment to be used and on reagents needed, contain an exact description of the analytical procedure including calibration, and information on the precision and accuracy of the procedure (as an example, see Fig. 10).

3.5.4
Analytical QA Measures

3.5.4.1
Control Analyses

Within analytical QA measures, special attention has to be drawn to the assurance of test results (validation [28]) (see also chapter 6 [29]). In addition to reducing analytical errors by technical developments and staff training, the production of reference materials and the participation in interlaboratory comparison tests, these measures include the requirement to perform control analyses regularly. A first "safety measure" of all analytical studies is to perform duplicate determinations with a subsequent statistical evaluation of the results. If the individual values diverge more than expected for the technique applied, "control analyses" are performed. If inadmissable and not directly explainable deviations occur during these control analyses, then, possibly, the use of this measuring and test equipment will be suspended pending further investigation (see Sect. 3.4.1).

A further measure for validating the analytical data is to perform control analyses on samples chosen at random and according to well-determined test procedures and/or recognized reference procedures. This means an increase in expenditure; this, however, should be considered necessary in relation to the possible harm from not doing it. The results of these control analyses are documented in the form of quality control charts (example see Fig. 11).

In this context it is certainly of interest to get an estimate of the expenditure needed for quality assurance. The total expenditure spent within the laboratories for the validation of analytical results and for their control, differs

Dokumentation pertaining to the QM-Analytical Laboratory	
Test instruction for the determination of .... (Element / Matrix)	Reference number: ....

Purpose: ....

Fundamentals: (short description of the chemical of physical fundamentals; reaction formulas if appropriate)

Reagents: ....

Operation: (complete description of the analytical procedure)

Titer determination resp. construction
of the analytical function: (complete description of the procedure
e. g. for titer determination of a titra-
ting solution or the construction of
the analytical function in the case of
photometric or spectrometric studies)

Admissible variation: (statement of the standard deviation of the repeatability
depending on the analyte content)

Reference Material: (indication of the reference material used for verification of
the procedure or for the construction of the analysis func-
tion)

Action in the case of deviations: (description of the procedure to follow, e. g. to
repeat analysis according to the operation in-
struction)

Person responsible for the study: (signature of the person in charge)

Drawn by: (signature)   Checked: (signature)

First issue: Validity:	Revision number: Date:

Fig. 10. Characteristics of an analytical procedure

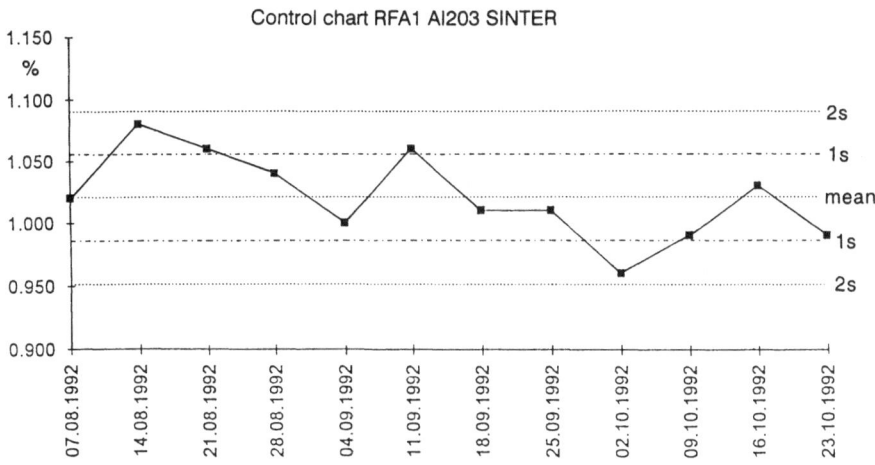

Fig. 11. Example of a control chart for analysis control: Determination of Al_2O_3 in a sintered material by X-Ray Fluorescence Spectrometry

from one laboratory to another; however, one may start from the assumption that it amounts to up to about 30% of the total laboratory expenditure in time and costs. From this fact it necessarily follows that analyses of comparable quality can only be obtained by the application of comparable systems and by the existence of corresponding experience.

3.5.4.2
Reference Materials

When considering the validation of analytical results, fundamental importance should be attributed to reference materials (RM) or certified reference materials (CRM) (see also chapter 8 [31]). Certified reference materials (CRM), i.e. materials with confirmed amounts of substance, are produced and distributed by internationally recognized organizations and institutions (see also chapter 7 [32]). CRMs used for quality assurance are kept available in the laboratories concerned. Results obtained using these materials are recorded in the department concerned. In analytical laboratories, according to the cited standard, reference materials are used for calibration ("calibration standards") as well as for controlling ("control samples for analyses"). However, the certified data alone do not guarantee the successful, i.e. correct application of reference materials. Depending on the material to be investigated or on the technique to be applied, a specialist assessment and a problem-related selection is needed. It follows, that a task-related application of either an (instrumental) analytical technique or the calibration standards still demands a professionally trained specialist.

3.5.4.3
Interlaboratory Studies

"Interlaboratory studies" (round robin analyses) are performed (see also chapter 9 [33]) in order to distinguish the quality of analytical techniques, e.g. in the framework of standardization, and during the preparation process of certified reference materials. After their statistical evaluation, the results of these studies serve as an evaluation basis. These cooperative investigations are also a necessary procedure to secure quality assurance in analytical laboratories.

An aim of these cooperations is that participants gain insight into their own performance and into the comparability of analytical results. The results of these

Table 2. Results of an interlaboratory study for the certification of a reference material (example: boron containing chromium-nickel steel)

	Laboratory means based on 4 determinations in m/m %										
Lab. Nr.	C	Si	Mn	P	S	Cr	Mo	Ni	B	Co	Cu	N
1	0.0148	0.5362	-	0.0248	0.0009	18.48	0.2330	10.24	0.8291	0.1407	0.1930	0.0183
2	0.0156	0.5410	1.451	0.0249	0.0010	18.48	0.2370	10.24	0.8375	0.1418	0.1956	0.0184
3	0.0157	0.5470	1.464	0.0252	0.0011	18.50	0.2391	10.25	0.8399	0.1430	0.1972	0.0185
4	0.0157	0.5495	1.465	0.0258	0.0011	18.52	0.2399	10.26	0.8470	0.1430	0.1995	0.0185
5	0.0158	0.5523	1.467	0.0262	0.0012	18.52	0.2425	10.31	0.8575	0.1431	0.2005	0.0187
6	0.0158	0.5543	1.467	0.0264	0.0012	18.56	0.2445	10.32	0.8729	0.1457	0.2018	0.0192
7	0.0162	0.5550	1.472	0.0265	0.0014	18.57	0.2452	10.32	0.8778	0.1465	0.2018	0.0194
8	0.0162	0.5675	1.474	0.0267	0.0015	18.61	0.2455	10.32	0.8870	0.1470	0.2020	0.0196
9	0.0163	0.5680	1.475	0.0268	0.0015	18.62	0.2456	10.34	0.9050	0.1482	0.2033	0.0196
10	0.0166	0.5749	1.475	0.0268	0.0015	18.62	0.2456	10.36	0.9102	0.1488	0.2038	0.0196
11	0.0167	0.5778	1.477	0.0270	0.0016	18.63	0.2458	10.36	0.9273	0.1488	0.2042	0.0197
12	0.0168	0.5791	1.479	0.0270	0.0017	18.63	0.2480	10.38	0.9283	0.1490	0.2050	0.0197
13	0.0171	0.5805	1.480	0.0274	0.0018	18.63	0.2508	10.39	0.9302	0.1490	0.2050	0.0203
14	0.0173	0.5835	1.480	0.0276	0.0018	18.64	0.2527	10.40	0.9344	0.1504	0.2058	0.0206
15	0.0178	0.5844	1.482	0.0281	0.0021	18.65	0.2535	10.40	0.9375	0.1513	0.2062	0.0207
16	0.0179	0.5870	1.483	0.0282	-	18.66	0.2542	10.40	0.9400	0.1548	0.2070	-
17	-	0.5880	1.486	0.0293	-	18.67	0.2548	10.42	-	0.1558	0.2095	-
18	-	0.5888	1.504	-	-	18.74	0.2560	10.44	-	-	0.2099	-
19	-	0.5940	1.526	-	-	10.77	0.2650	10.44	-	-	-	-
M_M	0.0164	0.5689	1.478	0.0267	0.0014	18.61	0.2473	10.35	0.8914	0.1475	0.2028	0.0194
s_M	0.0008	0.0181	0.016	0.0012	0.0004	0.08	0.0076	0.07	0.0401	0.0043	0.0045	0.0008

M_M = arithmetic mean of the laboratory means

s_M = mean standard deviation of the laboratory means

Certified values with respect to the statistical evaluation of the individual values

(in m/m %):

Lab. Nr.	C	Si	Mn	P	S	Cr	Mo	Ni	B	Co	Cu	N
M_M	0.016	0.569	1.48	0.027	0.0014	18.61	0.247	10.35	0.89	0.148	0.203	0.019
s_M	0.001	0.018	0.02	0.001	0.0004	0.08	0.008	0.07	0.04	0.005	0.005	0.001

interlaboratory rounds allow every participant to derive for himself the suitability of the applied analytical technique, his staff's professionality, and the reliability of his equipment, and, thus, among other things to prove his competence. The results obtained are available as documentation (for example, see Table 2).

3.5.4.4
Internal Quality Audits

Internal quality audits are performed with the aim of determining the actual state of the quality management system and of the course of quality assurance measures, and, to attain continuous progress and improvement. Therefore, the QM system is audited at least every 2 years. The working basis for the audits are the quality manual (QM) together with the relevant manuals, records and documentation.

Internal quality audits are planned and prepared by the person responsible for the QM of the company department, that is the person in charge of QA. However, they are performed by auditors not belonging to the department to be audited. Planning and working out the program for the audit is done with the assistance of specialists of the analytical laboratory. The auditors compile the final report including possible corrective measures that have been agreed upon. Carrying out the scheduled corrective and improvement measures is overseen and reviewed by repeated audits if necessary.

3.6
Process Capability and Machine Capability

Recently, the terms "process capability" and "machine capability" play a central role in the statistical evaluation of production and measurement processes. Process capability means that the production process (analytical process) constantly meets the quality requirements, while machine capability provides information as to whether a device or a procedure is operational with recognizable uniformity within preset tolerance limits.

Process capability is an index number, which informs if a process can be run within the required limits. The variation and the accuracy of analytical results exert a significant influence on this index number, which should not be less than unity. Process capability as well as machine capability (> 1.3) can be improved by smaller standard deviations and by the elimination of systematic errors. Reporting the capability indices with time is an appropriate means of proving and assuring analytical quality and, therefore, it also contributes to ensuring the quality of processes related to analysis [36].

While machine capability is a short-term observation, to prove process capability implies a long-term study. In both cases, the variation of the (analytical) process, expressed as standard deviation, play an important role in the calculation of the indices (see Figs. 12 and 13) [37, 38]. Here, the approach deviates from the pure analytical process: the variation should be in appropriate

- only random variation
- normally distributed

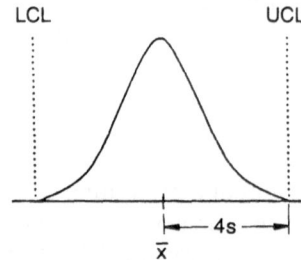

Minimal requirement: $\bar{x} \pm 4s$ specification
$$= 99.994 \%$$

$$\Longrightarrow \quad \boxed{c_m = \frac{UCL - LCL}{6s} \geq 1.33}$$

resp. considering the mean value position:

$$\boxed{c_{mk} = min \left(\frac{UCL - \bar{x}}{3s} \; ; \; \frac{\bar{x} - LCL}{3s} \right) \geq 1.33}$$

Fig. 12. Calculation of machine capability

- long-term study
- cooperation of personnel, machinery,
  raw material, methods and work environment

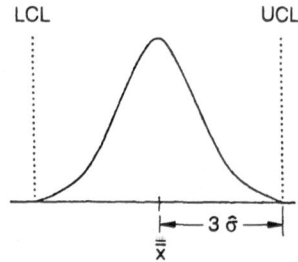

Minimal requirement: $\bar{\bar{x}} \pm 3 \, \hat{\sigma}$ specification
$$= 99.73 \%$$

$$\Longrightarrow \quad \boxed{c_p = \frac{UCL - LCL}{6 \, \hat{\sigma}} \geq 1.0}$$

$$= estimate \quad \hat{\sigma} = \frac{\bar{R}}{d_2} \quad or \quad \hat{\sigma} = \frac{\bar{s}}{c_4}$$

d2, c4 are depending on the sampling size.
Process coefficient:

$$\boxed{c_{pk} = min \left(\frac{UCL - \bar{\bar{x}}}{3 \, \hat{\sigma}} \; ; \; \frac{\bar{\bar{x}} - LCL}{3 \, \hat{\sigma}} \right) \geq 1.0}$$

Fig. 13. Calculation of process capability

relation to a tolerance that is determined to be the result of the technical process by defining an upper and lower limit. The process capability index informs us whether it is possible to run the specific process in the required way.

In this context, the mean value of the measured test parameter gains a special significance. The process is only operable if this mean equals a target value over a given period of time. Machine capability determines the performance of a machine or a measuring device with regard to its keeping within defined tolerance limits. In this case, requirements are higher because a variation of 4 s (99.994% of all test values) is used for the calculation. Optimum performance of the machine or measuring device has to be proven by performing a great number of measurements, over a short period. Systematic errors have to be eliminated prior to study process capability and machine capability.

3.7
Certification of Quality Management Systems and Accreditation of Anaytical Laboratories

At home and abroad, an increasing number of purchasers of industrial products expect their suppliers not only to run an effective quality management system but also to have system effectiveness proved by a neutral body. Therefore, in the interests of industry and commerce, organizations which issue such certificates were inaugurated. In this way, in addition to the certification of quality management systems, a need became apparent for the accreditation of testing laboratories, i.e. for chemical analytical laboratories, too. As a basis for these activities there exists the already mentioned European series of standards EN 45000 according to which accreditation of a testing laboratory means the formal recognition of the competence of a testing laboratory to effect special tests or types of tests [39].

An accreditation procedure starts with submitting an application to an accreditation body; this application lays down to what extent the laboratory wants to be accredited. At this first stage, general information on the testing laboratory is gathered, e.g. on staff and equipment. Already at this point, it should be possible to present a quality manual. In the further course of the procedure, an assessor examines the staff's qualifications, the instrumental equipment used to perform application-related procedures, the quality of the test facilities and, especially, the competent conduct of analytical studies (see chapter 10 [40]). This examination is done during an on-site assessment, i.e. by a visit by either the auditor or several specialist assessors. If the result of the assessment procedure is favourable, i.e. the laboratory meets the requirements for accreditation, then an accreditation certificate is issued.

However, accreditation is only granted if the laboratory is under constant supervision. Therefore, at regular or irregular intervals, the laboratories are inspected again, either formally or by a visit. As a rule, an accreditation is for a limited time. When this expires (for example after five years), a new accreditation is possible if the laboratory so wishes. To obtain accreditation, of course, a free

decision of the individual laboratory; there is no stringent governmental guideline. The market, alone, is the determining factor. Many laboratories, however,
have recognized the increase in competitiveness accreditation gives and take part
in the procedure in order to demonstrate their analytical quality.

The accreditation of testing laboratories in the statutorily non-regulated
area is further accompanied by a considerable economic aspect. Assuming
that, in the future, all testing laboratories within the common market are
accredited by their national accreditation bodies following the same criteria
(EN 45002), then no compelling reason exists any longer to carry out the same
test on product conformity at the customer's site. This (theoretically) means
saving half of the test costs.

Advantages resulting from the accreditation of an analytical laboratory can
be summarized as follows:

1. A qualified fulfillment of customer expectations for the effectiveness of the
   QM system,
2. A stronger position when auditing a company's QM system,
3. Harmonization of commercial arbitration analysis (here, for example, a fundamental difference exists between mechanical-technological and metal-
   lurgical testing),
4. Proven competence with regard, to the execution of assignments from outside of the company,
5. Qualified participation in interlaboratory studies on the production certified reference materials (CRMs).

This last mentioned aspect represents the best and most economic opportunity for selfcontrol and an examination of one's own performance. Furthermore,
it offers the opportunity to contribute with one's competence to the CRM-
procedure, and, to exert an influence on the CRM-process.

At this point, it is impossible to go into details on the complex multi-level
cooperation of all the different organizatons working in the field of certification and accreditation. One has to refer to the corresponding literature (see
chapter 12 [41]).

The above-mentioned clearly elucidates the importance QA has already
gained or will develop in an increasing manner for analytical chemistry. Due
to the EC internal market, independent certificates on a functioning QM
system will gain importance for every company – and, therefore, for analytical
laboratories, too. Only by strict quality-assured practices will it be possible to
minimize quality problems arising in production and the use of products and,
as well, when providing services.

3.8
References

1. Koch KH (1990) (Oktober) Kontrolle, S 80/87
2. Funk W, Dammann V, Donnevert G (1992) Qualitätssicherung in der Analytischen Chemie. VCH Verlagsgesellschaft mbH, Weinheim
3. Theuer A (12/1990) LABO, S 7/13
4. DIN 55350 (Teil 11): Begriffe der Qualitätssicherung und Statistik – Grundbegriffe der Qualitätsicherung
5. DIN/ISO 9004: Qualitätsmanagement und Elemente eines Qualitätssicherungssystems, Leitfaden
6. Rehm S (1991) (November) Kontrolle, S 71–75
7. Qualitätsstrategie – Die Grundlagen. Hrsg. Hoesch AG, Dortmund, 4. Aufl., Januar 1992
8. Bläsing JP (1992) Das qualitätsbewußte Unternehmen. Steinbeis-Stiftung für Wirtschaftsförderung, Stuttgart
9. Walton M (9190) Deming Management at Work. G.P. Putnam's Sons. New York
10. Crosby PB (1986) Qualität ist machbar. McGraw-Hill Book Company GmbH, Hamburg
11. Taguchi S, Byrne D (1988) Die Taguchi-Methode des Parameter Designs. Praxishandbuch Qualitätssicherung (Herausgeber Bläsing) Band 4. GFMT-Verlag, München
12. Bläsing JP (1988) Qualitätspolitik legt die Leitlinien der unternehmerischen Qualitätssicherung und die spezielle Qualitätsverantwortung fest. Praxishandbuch Qualitätssicherung (Herausgeber Bläsing) Band 4. GFMT-Verlag, München
13. Zink JK (1989) (Herausgeber): Qualität als Managementaufgabe. Total Quality Management. Verlag moderne Industrie, Landsberg
14. Thierig D, Thieman E (1983) Arch Eisenhüttenwes 54, S 301/305
15. Czabon V (1992) Fresenius' J Anal Chem 342, S 760/763
16. Hartmann E (1992) Fresenius' J Anal Chem 342, S 764/768
17. Qualitätssicherungshandbuch und Verfahrensanweisungen DGQ-Schrift Nr. 12–62, 2. Auflage, 1991, Deutsche Gesellschaft für Qualität e.V. (DGQ), Frankfurt/M; Beuth Verlag GmbH, Berlin
18. Böshagen U see Chap. 2
19. DIN ISO 9000: Leitfaden zur Auswahl und Anwendung der Normen zu Qualitätsmanagement, Elementen eines Qualitätssicherungssystems und zu Qualitätsnachweisstufen. DIN/ISO 9001: Qualitätssicherungs-Nachweisstufe für Entwicklung und Konstruktion, Produktion, Montage und Kundendienst. DIN/ISO 9002: Qualitätssicherung-Nachweisstufe für Produktion und Montage. DIN/ISO 9003: Qualitätssicherungs-Nachweisstufe für Endprüfungen
20. Danzer K see Chap. 5, p 105–134
21. Merz W, Weberruß U, Wittlinger R (1992) Fresenius' J Anal Chem 342, S 779/782
22. ISO IEC Guide 43: Development and operation of laboratory proficiency testing
23. ISO IEC Guide 45: Guidelines for the presentation of test results
24. DIN ISO 10012, Teil 1 (Entwurf): Forderungen an die Qualitätssicherungen von Meßmitteln – Meßmittelmanagement
25. VDI VDE/DGQ-Richtlinien, Blatt 1 (1980) Prüfanweisungen zur Prüfmittelüberwachung/Einführung
26. Werner W (1992) Fresenius' J Anal Chem 342, S 783/786
27. Wegscheider W „Richtige Probenahme, Voraussetzung für richtige Analysen": s. S. 61 see Chap. 4, p
28. Ebel S (1992) Fresenius' J Anal Chem 342, S 769/778
29. Wegscheider W see Chap. 6, p 135–158
30. Koch KH (1982) Arch Eisenhüttenwes 53, 97–100
31. Griepink B „Referenzmaterialien für die Qualitätssicherung"; s. S. 157
32. De Bièvre P see Chap 7, p 159–193
33. Cofino WP see Chap 9, p 209–217

34. DIN ISO 5725: Präzision von Meßwerten – Ermittlung der Wiederhol- und Vergleichsprä-
    zision von festgelegten Meßverfahren durch Ringversuche
35. Grasserbauer M, Pfannhauser W, Wegscheider W (1987) ÖChemZ 88, 130–123
36. Thierig D (1991) Stahl und Eisen 111, Nr. 10, S 83/87
37. Ford AG, Köln (1985) Statistische Prozeßregelung, Qual Cont EV 880b
38. Ford AG, Köln (1988) Q 101, Qualitätssystem, Richtlinie
39. Staats G, Tröbs V (1992) CLB Chem Lab Biotechn 43, H 4, S 194 197, H 6 S 314 318
40. Mechelke GJ see Chap 10, p 219–227
41. Günzler H see Chap 12, p 241–246

Chapter 4

Proper Sampling: A Precondition for Accurate Analyses

Wolfhard Wegscheider

Abstract

There can be no discussion of sampling without an unambiguous problem definition and a clear formulation of the analytical task. The systematics of the series EN 45000 stipulates that important sampling can only be carried out by surveillance laboratories which are authorized to draw conclusions from single samples to the whole.

In those cases where the analytical sample represents the entire population, sampling is inapplicable as a selection process. This is true for very small samples; the so-called integration error then becomes zero. In general, inhomogeneities resulting from the discrete structure of material, from the persistence and the periodicity of a process (autocorrelation), contribute to this integration error.

Strategies for the reduction of these error components are briefly discussed.

4.1
Sampling Within the Analytical Process

Analytical quality, like quality in general [1], can only be measured through customer satisfaction. Therefore, a decisive influence on the success of an analytical laboratory is not only exerted by precision and accuracy, but there are also numerous other criteria, such as: price, customer relations, observance of deadlines and, of course, sampling. Regardless of whether sampling is done by the accredited laboratory, by the customer himself or by a third party, the benefit the customer may derive from the analytical data is inseparably connected with sampling.

However, the analyst should take into account that EN 45001 explicitly forbids testing laboratories to draw conclusions from analytical data; thus specialist sampling is implicitly confined to inspection and certification bodies. Depending on the problem, the "whole" may have many different meanings; in each case, the total object of an investigation, called the "population" in the classical terminology of sampling statistics, has to be considered; hence, for example, a study field, a study period, a lot, a batch, a lorry, a geologically interesting stratigraphical horizon.

This embodiment of sampling competence among the tasks of inspection bodies, of course, does not exclude the important role that analytical aspects play within the sampling process, especially those aspects with regard to stability and the original binding state (speciation) of the analyte. Rather it means that it is not certain that everyone in the laboratory who is capable of determining a certain analyte in a given matrix, is also capable of carrying out a technically correct and representative sampling which allows one to draw far-reaching conclusions about the population.

Sampling has to be regarded as the material connection between analyst and contractor, as the link between analytical result and problem: the better this link, the more likely the laboratory will be able to offer real problem solving and not only (more or less) accurate data. The final result of analysis includes the uncertainty contribution arising from the sampling process as well as from the measurement itself [3]; the same is true in the case of systematic errors because there is no difference whether the systematic error results from sampling or from analysis. Therefore, sampling needs the same meticulousness, attention, and accuracy as analysis. If the sample is erroneous, i.e. affected by a systematic error, then not even the best analyst can recognize this error – let alone correct it.

While sampling is directly relevant for commercial and technological problems because otherwise an actual transaction or a development under consideration cannot be carried out, for other areas such as environmental analysis, food control etc., it is in no way of minor significance, however, there, less automatic checking is possible. In this case, there is no direct feedback (and, therefore, also no direct correction) of a questionable sampling strategy: At best, many of these data are statistically or graphically processed without drawing any direct consequences even elementary ones. This principle seems to be especially true in the case of "normal" values, at any rate, as long as they stay below a critical boundary (limit value).

In many cases, the so-called plausibility control provides a last regulatory mechanism before releasing data and, in part, is falsely considered to substitute for quality assurance measures in the laboratory. It is just in these cases which are of special public interest that the question of proper sampling has to be addressed as critically as the question of accurate analytical resuts.

4.2
There Is No "Correct" Sampling Without A Clear Problem Definition!

It is a fundamental problem that there can be no proper sampling without a clear problem formulation or, in other words, if no questions are asked, there can be no answers, only data. This distinction is equivalent to the distinction between data and information often made in common parlance. Here implicit questions are not admissable because then, contractor and laboratory might put different interpretations on them. Therefore, in the relevant standards, great emphasis is put on regular and sufficient communication between contractor and laboratory; this should be done in writing for reasons of traceability. At this point, a regulation following GLP [4], which stipulates a preceding written agreement between contractor and laboratory may be considered as ideal. Hence, for an accredited laboratory there should be operating procedures which describe how to plan a sampling process. Planning, like sampling itself, will often be a very complex and therefore expensive process because one should not start from the assumption that inexperienced or "cheap" staff may be employed. In all those cases where sampling is not clearly defined by a standard, and thus the laboratory is held responsible for the meaningfulness of the procedure and the representativeness of sampling, planning of sampling and its handling is in the hands of the laboratory management. Where standard specifications exist, training and constant supervision of staff entrusted with sampling will be an important and time intensive activity of the executives of an accredited laboratory, particularly since sampling is not easily covered by classical methods of quality assurance (e.g., control charts, statistics, spiking techniques, etc.). Furthermore, great importance should be attributed to the use of best equipment in order to make well-conducted sampling possible even under unfavorable external conditions (weather!). Frequently, a specially adapted vehicle is put at the disposal of each sampling team enabling on-site sample conservation in professional manner.

The alternative, i.e. not to be concerned with problem definition and, thus, also with sampling, is, in the long run, not a healthy attitude for chemical laboratories – and hence for the role of analytical chemistry in society – because they will be easily forced into the awkward situation of being the cause of unnecessary expense. Admittedly, it is the customer who is unable to utilize the data, but, at some time or other, the customer will realize this and then will attribute joint responsibility to the laboratory.

4.3
Managing Without Sampling?

Sampling is an error generating step or – in technical terms – an extra, often considerable contribution to the measurement uncertainty [5]. Thus, it has to be examined if cases exist where sampling may be omitted. This is possible whenever the whole sample (population) is so small that it can be analysed in its totality. Is it about an fault in a masterpiece, e.g. a discolouration at the left corner of Mona Lisa's mouth, the sample may only consist of the discoloured part of the corner of the mouth, which then, probably, will be analyzed in total using, hopefully, a non-destructive method. Analogous examples exist in forensic chemistry (flakes of paint after motorcar accidents, hair in the case of violent crimes, etc.) and also when searching for sources of error that are causing the defect of an individual product. In such cases, problem definition ensures that only a well-defined population will be considered, AND, that this population, as a laboratory sample, has to be analyzed in its entirety. The latter is only possible for analytical methods not affected by remaining inhomogeneities; hence, if the spatial and/or time-dependent resolution of the analytical method is much coarser than the inhomogeneity (regions) and, therefore, the inhomogeneity principally cannot be observed by the analyst.

4.4
Planning Sampling Procedures

Thus, in connection with any planning of sampling procedures, there are two crucial aspects:

a) the exact description of the item to be studied in its whole *spatial and temporal extension*; this should be perfectly clear from the problem definition, and,
b) the estimation (experience, pre-information, test measurement) of the *inhomogeneity* of the population; whoever performs sampling has to be informed on the approximate degree of inhomogeneity, and, here again, its spatial and time dependence has to be taken into consideration.

Homogeneous samples do not represent any problem for sampling because each part of this item constitutes a utilizable representation of the whole. This, of course, always with the assumption that the analyte is not modified by the sampling technique. Despite its eminent practical significance, this fact, as well as the question of analyte stabilization prior to analysis will not be discussed further here. It is obvious that the most important error remaining in sampling stems from the fact, that homogeneity is presumed even in those cases were it does not exist at all. The essential question to be answered is always whether the analytical process will be influenced by the inhomogeneity or not, because, in a strict sense, on the level of individual atoms (for example in surface analysis) homogeneity is an idealized concept and not realized in nature.

A simple example may illustrate this: research on forest decline can be performed via satellite or after sampling of needles or leaves. While the use of satellite demands a homogeneity of the order of the resolution of its optical system (approx. 10×10 m), a direct analysis of needles to determine sulfur by for example X-ray fluorescence spectroscopy asks for a homogeneity within the order of $10\,\mu$m, because the depth of emerging characteristic X-ray emission of sulfur probably will not be greater than this. The decision on whether preference has to be given to the one system (satellite) or to the other system (XRF) can certainly not be taken without a detailed problem discussion.

Thus, the better the spatial and temporal resolution of the analytical system, the more detailed can information be obtained by corresponding effort, but, the less integral information (mean values) can be provided by the analytical system. Which one of these two aspects, i.e. to obtain good resolution or to get good average values, will be more important for a certain problem depends exclusively on the problem definition.

4.5
Aspects of Measurement Uncertainty Caused by Sampling

Two totally different components contribute to the often important uncertainty entailed in the sampling process and which finally contributes to the uncertainty of the analytical data:

a) uncertainties regarding to homogeneity assumptions, and,
b) uncertainties associated with the sampling process that is deduced from these homogeneity assumptions.

Assumptions regarding homogeneity more or less conform to reality and are formulated as a mathematical model; one presumes, that this model is valid for the whole population. These assumptions may be: expected analyte content, particle size and distribution, degree of segregation, grade of homogeneity obtained by grinding or other crushing processes. For continuous processes the assumptions include information related to periodicity, time constants, stationarity and to the type of the autocorrelation function. In this context, one has to refer to the literature [6, 7] for more detailed information. The fact remains that, in the end, these data are only known approximately, and, hence, this contributes to the final result as model errors.

However, the uncertainties arising from b), i.e. those resulting from the realization of the model according to a), should be examined more closely in order to make accessible to the reader the diversity of considerations to be input in the planning of a sampling process and carrying it out. The following comments are based on work of the doyen of sampling, Pierre Gy [8] and are related to a single measurement. The advantage is that all uncertainties manifest themselves as "errors" and this latter term is anchored better in analytical chemistry than "measurement uncertainty". As shown in Fig. 1 of reference [3], numerous elements contribute to the total error. Here, like in analytical chemistry in general, it cannot be assumed that all activities in the field, i.e., on-site, may be classified as being part of sampling and all activites thereafter as belonging to analysis. Very often, sample mass (volume) is reduced in the laboratory by a corresponding selection and homogeneization process; this, from its systematics, naturally corresponds to a sampling step. Globally, it can be written:

$$s_{tot}^2 = s_{sa}^2 + s_{anal}^2$$

with s_{tot}^2 being the total (quadratic) error, s_{sa}^2 being the sampling error and s_{anal}^2 being the analytical error which not only takes into account the measurement itself, but also those sample preparation steps which are not concerned with sample reduction and homogenization. To understand the meaning of sampling error, it is important to consider that sample reduction (i.e. the selection of a portion out of a large quantity) and homogenization generally has to be carried out in several successive steps; though each of these furnishes a different contribution to the total error, from their structure, they may be treated in an analogous way. For n successive selection and reduction steps, hence

$$s_{sa}^2 = \sum_{i=1}^{n} s_{sa,i}^2,$$

but, here, the discussion is based on $n = 1$.

The sampling error itself is a combination of an error contribution related to sub-sample selection (selection error) and one related to sample preparation for a subsequent sampling phase or for analysis. Such preparative steps do not really serve as selection process, but, by the necessary manipulations, sample elements can be subjected to forces which may lead to a selection in the sense that they give preference to certain sample components, for example by a different density during (attempted) homogenization, or by different hardnesses of individual particles when crushing the sample. The selection error, s_{sel}^2, itself, can not be estimated directly; it can only be assessed after an elemental breakdown into several steps. Thus, first we define:

$$s_{sa}^2 = s_{sel}^2 + s_{prep}^2 = s_{int}^2 + s_{mat}^2 + s_{prep}^2$$

The index "int" denotes the so-called integration error, related to the selection process, and the index "mat" (for materialization) denotes the error made during identification, delimitation and separation of the sub-samples.

4.5.1
Integration Error

The integration error itself takes into account the inhomogeneity of the material, i.e. the error contribution

- related to the discrete structure of the material (being either of the oder of magnitude of the particle size in the case of solids, or of ions, atoms, and molecules in the case of gases and liquids), and that
- which results from the continuous nature of the process that has supplied the material, and
- that error contribution which is determined by the periodicity of the process.

Hence:

$$s_{int}^2 = s_{discr}^2 + s_{contin}^2 + s_{period}^2.$$

Proper sampling requires methods of keeping all three components as small as possible, but, the strategy employed will be different for each of these components. The first component, i.e., the one attributed to the discrete structure of material, s_{discr}^2, can be reduced by the following measures: large quantity of sample, small particle size, minimal segregation (good homogenization) and small mass of each sub-sample.

Error contributions based on the continuous nature of a process, s_{contin}^2, can be reduced by repeated sampling of small increments, thus by small sub-samples sampled in short time intervals and small spatial increments. In a borderline case, for gases and liquids, this may lead to a continuous measurement. However, it is of fundamental significance to recognize that average results can never be obtained better than dictated by the process variability itself.

For the third type of error contribution based on inhomogeneity, i.e. those from periodic fluctuations, s^2_{period}, it has to be distinguished whether they are more or less important than the two types discussed previously (those caused by the discrete structure of material and those caused by the continuous nature of the process). If the contribution resulting from periodic fluctuations is relatively small, sampling can be based on a systematic selection of sub-samples (for example at constant time intervals). If the fluctuation amplitude is large thus overshadowing the two other contributions, stratified sampling should be applied. Large and small periods should be sampled separately and their combination should only be done mathematically, and not physically. The principle of stratified sampling is explained in Ref. [2].

4.5.2
Materialization Error

While the integration error of sampling will never be zero, except if one takes into account – as shown above – the whole population, this is entirely possible for the materialization error, s^2_{mat}, and, of course, this state of affairs has to be strived for. With regard to the development and assessment of sampling tools and devices, it is also useful to discuss the materialization error components separately. These are:

- the delimitation error of an aliquot,
- the extraction error of an aliquot, and,
- the preparation error of an aliquot.

The delimitation error occuring at sub-sample selection due to the sampling device, is the relative deviation between the theoretically correct volume and the actual volume sampled. It can be caused by a wrong geometry of that part of the sampling device that comes into direct contact with the (total) sample, instrument operation at the wrong speed, by a misinterpretation, by a design flaw, or by poor maintenance of the instrument.

The extraction error of a sub-sample, in the majority of cases is due to technical reasons. Extraction error denotes that error leading to a preferential extraction of a certain component of the sub-sample in spite of correct delimitation within the total sample material (in the total continuous flow). In this case, the problem encountered is that a selective and differential interaction takes place between the instrument and part of the sample material. As always in the case of systematic errors, here too, there are certain general guidelines and experience to eliminate this error.

Preparation errors leading to distortions may appear at several steps: during extraction, sample transport, or sample preparation. The analyst is familiar with classical elements of preparation error. These may be: contamination of the sample material, loss of material or analyte (for example through adsorption or degradation), changes in the material composition of the sample (for example by an alteration of the water content, by an oxidation of certain sample components), or by a loss during transport of especially large or small sized grains. Of course, unintentional errors (wrong labelling, erroneous mixing of different samples, use of unsuitable or contaminated devices) or intentional manipulation may render the sample useless even before analysis begins.

Even if this present discussion does not give any working instructions for the practitioner, it should be perfectly obvious that sampling is a difficult subject because its successful realization demands an extensive theoretical understanding of the problem and of the sampling process, as well as a specialist translation into practice while using the most appropriate sampling devices.

4.6
Conclusions

The great significance of sampling is explained by the fact that sampling constitutes the link between analytical data and a physical problem definition. As a rule, sampling generates a substantial contribution to the total measurement uncertainty. This contribution should be kept very small and technical competence is needed to quantify the uncertainty due to sampling. In many cases, at present, this requirement is not met by testing laboratories.

In the framework of international accreditation guidelines, it is not unambiguously defined whether sampling is confined to inspection laboratories authorized to draw appropriate far-reaching conclusions from extensive data material. In any case, the differentiation made in the EN 45 000 series between testing and inspection laboratories has to be interpreted in the way that testing competence (analysis) does not automatically include the qualification for competent sampling. Apart from suitable equipment, sufficient know-how is a principal requirement for proper sampling. Viewed from the systematics of metrology, it would be sensible to urge that sampling should be restricted only to those institutions competent to draw appropriate conclusions from a subsample, i. e. from single analytical results to the whole. In a nutshell: without adequate sampling there can be no valid conclusions.

4.7
References

1. DIN 55350, Begriffe der Qualitätssicherung und Statistik
2. Cochran WG (1963) Sampling Techniques 2. Aufl., Wiley, New York
3. Wegscheider W Chapter 6 in this volume
4. The OECD Principles of Good Laboratory Practice (1992) Environ Monograph No. 45, OCDE/GD (92) 32, Paris
5. BIPM/IEC/IFCC/ISO/IUPAC/IUPAP/OIML (1993) Guide to the Expression of Uncertainty in Measurement, ISBN 92-67-10188-9, Genf

6. Kateman G, Pijpers FW (1981) Quality Control in Analytical Chemistry, Cap. 2, Wiley, New York
7. Kateman G (1984) "Sampling". In: Chemometrics. Mathematics and Statistics in Chemistry. Kowalski BR(ed) NATO ASI Series C 138, Reidel, Dordrecht
8. Gy P (1992) Sampling of Heterogeneous and Dynamic Material Systems. Elsevier, Amsterdam

Chapter 5

Significance of Statistics in Quality Assurance

Klaus Danzer

Quality features cannot be reproduced with any arbitrary precision but only within certain tolerance limits. These depend on the process to be controlled as well as on the test method applied and on the (economically acceptable) testing and routine costs.

In the same way, analytical measurements for quality control are fundamentally subject to error – various types of error may appear and these can be influenced to varying degrees.

5.1
Types of Errors Associated With Analytical Measurements

Even when instrument reading are sufficiently accurate, repeated measurements of a sample lead, in general, to measurements which deviate by varying amounts from each other and from the true value of the sample. According to their character and magnitude, the following types of errors can be distinguished:

1. *Random errors* which manifest themselves as a distribution of the results of measurement obtained by repeated determinations around a mean of the sample and this variation is randomly distributed to higher and lower values. Random errors determine the *reproducibility* of measurements and therefore their *precision*.
2. *Systematic errors* displace the results of measurement one-sidedly to higher or lower values thus leading to incorrect results. In contrast to random errors, it is possible to avoid or eliminate systematic errors if their causes are known. Their existence and magnitude characterize the *accuracy* of a result of measurement.
3. *Outliers*, characterized as random errors which, however, have to be eliminated for the reason of their large deviation, so that the mean will not be distorted.
4. *Gross errors* which are generated by human mistakes or by instrumental and mathematical error sources, and, which, depending on whether they are short-term or long-term effects, may have systematic or random character. Often, it is easy to perceive and to correct for them. They will not play any role in the following discussion.

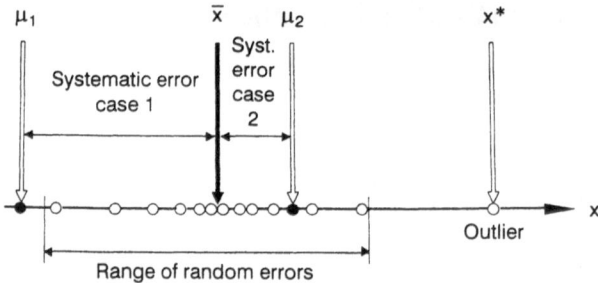

Fig. 1. Schematic representation of systematic and random errors

As well as the difference between systematic and random errors, the character of outliers is also clarified further by Fig. 1. The normal distribution of the measurement values determines the range of random errors. Measurement errors outside this range are denominated as outliers. Systematic errors are determined by the relation of the true value μ and the mean value $\bar{x}$ of the measurements and, in general, can only be recognized if they go beyond the range of random errors on one side.

Conventionally, a measurement result is said to be correct if the true value lies within the confidence interval (range of random errors) of the observed mean (μ_2; case 2 in Fig. 1). If the true value is located outside this range (μ_1; case 1 in Fig. 1), then the result is incorrect.

It is not always possible to discern strictly between random and systematic errors, especially as the latter are defined by random errors. The total error of an analytical determination, the *analytical error*, is, according to the laws of error propagation, composed of error contributions resulting from the measurement as well as from other steps of the analytical process [1, 2]. These errors include, as well as random, ones systematic contributions in some cases.

The relevance of systeamtic errors and thus the accuracy of analytical results, as well as the significance of outliers, is checked by statistical tests. These are based on distributions and variations which will be characterized more closely later on.

5.2
Systematic Errors

Systematic errors, in addition to random errors, may occur at all steps of the analytical process and, to be precise, for example during

- *sampling* by incorrect preferment of individual sample fractions,
- *sample preparation* by incomplete dissolution, separation or enrichment operations,
- *measurement* caused by concurrent reactions or incomplete reaction processes in the case of chemical principles, resp. by instrumental errors or

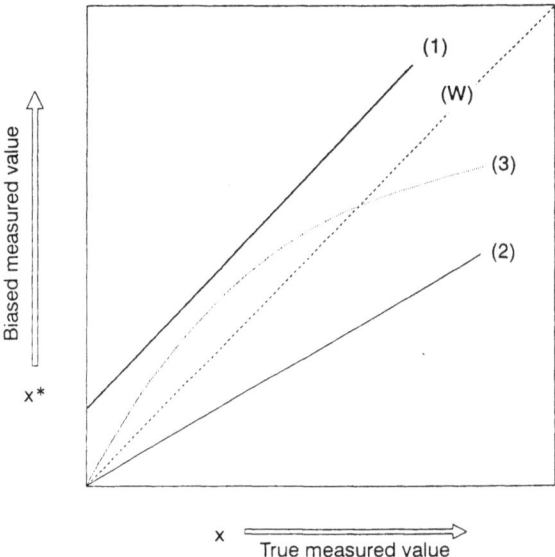

Fig. 2. Effect of systematic errors on measurement values; (W) ideal recovery (line of true measurements)

wrong adjustment in the case of physical methods. A frequently encountered reason for the occurrence of systematic errors is erroneous calibration due to unsuitable calibration standards, matrix effects, or insufficient methodical or theoretical foundation.

Even *data evaluation*, often thought free of errors to a large extent, can generate systematic errors by reason of incorrect or incomplete algorithms.

According to their influence on the measurand, one can distinguish (Fig. 2):

1. *Additive errors* altering the measurement values by a constant value. Instead of the true value x one measures the falsifed value

$$x^* = x + a \tag{1}$$

Reasons may be, for example, not recognized blanks.

2. *Multiplicative errors*, proportional to the measurement value and changing the slope of the calibration curve and thus sensitivity, because, instead of the true value x the falsified value

$$x^* = b \cdot x \tag{2}$$

is measured. They are often generated by erroneous calibration factors.

3. *Nonlinear, response depending errors* causing an incorrect value

$$x^* = x^k \tag{3}$$

to be measured instead of the true value; as a result, the calibration linear relationship between measurement and analytical value is lost. In Atomic Emission Spectroscopy, self-absorption of resonance signals generates such an effect.

Several of the mentioned types of systeamtic errors frequently occur together. In practice, they are best recognized by the determination of x^* as a function of the true value x, thus, by analysis of certified reference materials (CRMs). If these standards are not available the use of independent analytical methods or of a balancing study may provide information about systematic errors [3, 4]. In simple cases, it is also possible, to eliminate the unknown true value x through appropriate variation of the test portions or standard additions, and, to determine the parameters, a, b, and k by means of statistical mathematical methods.

In the framework of quality assurance, the use of reference materials is indispensable for the validation of analytical results (refer to chapter 8).

5.3
Random Errors

Repeated measurements performed with one and the same sample always result in random variations, even with careful control of constant experimental conditions. These are caused by measurement-related technical facts (e.g. noise of radiation and voltage sources), sample characteristics (e.g. inhomogeneities of solids), as well as chemical or physical procedure-specific effects. Random errors can be minimized, but basically not eliminated, i.e. they appear to be following a natural law. Therefore, they may be characterized by mathematics, namely, by the laws of probability and statistics.

5.3.1
Frequency Distributions of Measurement Values

Classifying varying measurement values by their magnitude does not, as a rule, give rise to uniform distribution over the whole variation range, but, results in a relative cumulation around the mean value, as shown, for instance, by the bar graph in Fig. 3.

Increasing the number of repeated measurements to infinity while decreasing more and more the width of classes (bars), normally leads to a symmetrical bell-shaped distribution of the measurement values, which is called *Gaussian* or *normal distribution*.

The frequency density $p(x)$, shown on the y-axis in Fig. 4a, is defined by the relation

$$p(x) = \frac{1}{\sigma \sqrt{2\pi}} \exp\left[-\frac{(x-\mu)^2}{2\sigma^2}\right] \tag{4a}$$

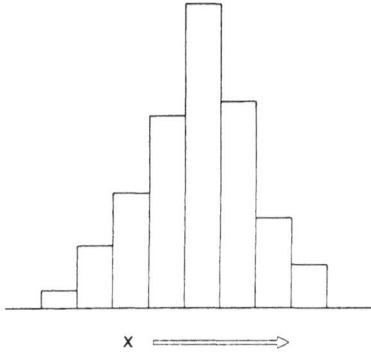

Fig. 3. Example of a frequency distribution of measurement values

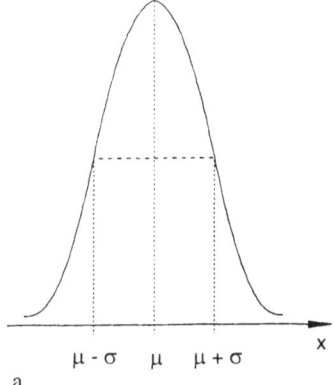

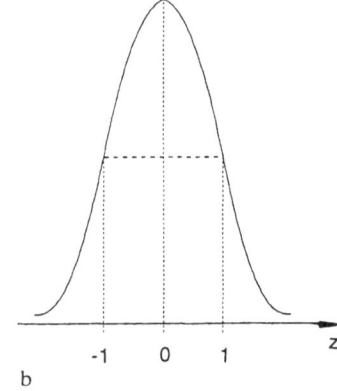

a   b

Fig. 4. Gaussian distribution (normal distribution, a) and standard normal distribution (b)

with the parameters μ being the maximum (*mean value*) and σ being half the distance of the inflexion points (*standard deviation*).

Using standardized values (Fig. 4b) with $z = (x - \mu)/\sigma$ leads to

$$p(z) = \frac{1}{\sqrt{2\pi}} \exp\left(-\tfrac{1}{2} z^2\right) \tag{4b}$$

In analytical practice, a population is studied by sub-sampling, i. e. the relation described by Eqs. (4a) and (4b) is only aproximate, and, instead of the parameters μ and σ, their estimates, $\bar{x}$ and s, are determined as arithmetic means

$$\bar{x} = 1/n \sum_{i=1}^{n} x_i \tag{5}$$

and standard deviation[1]

$$s = \sqrt{\sum_{i=1}^{n} (x_i - \bar{x})^2/(n-1)} \tag{6}$$

for n individual measurements, respectively. For real situations, the normal distribution $N(\sigma, \mu)$ is approximated by the t-distribution (STUDENT-distribution) $N_t(s, \bar{x}, n)$ which takes into account the number of individual measurements n or the number of degrees of freedom f.

Of further interest for analytical chemistry are the logarithmic normal distribution, the Poisson distribution as well as the F- and the χ^2 distributions. Like the t-distribution, the latter are important test distributions for statistical significance tests and for the calculation of confidence intervals (confidence ranges, uncertainty limits, tolerance and warning limits). These distributions are extensively studied in [5, 6].

Quality Assurance in analytical chemistry relies to a major extent on the distribution of mean values, which are in general normally distributed, and, to be precise, according to the Central Limit Theorem of statistics [6] even in the case of a non-normal distribution of the related individual values.

5.3.2
Error Propagation

The total error of an analytical process is combined by the error contributions of all steps of the analytical procedure (e. g. sampling, sample preparation, dissolution, separation, measurement) and of single operations (e. g. difference and comparison measurements). The combination of individual errors to give the total error, is determined partly by statistics and partly by functional relationships of the form $x = f(x_1, x_2, \ldots, x_n)$. In the last case, the total error can be estimated according to the general law of error propagation for independant variables $x_1, x_2, \ldots, x_n$

$$\sigma_{\bar{x}}^2 = \left(\frac{\partial f}{\partial x_1}\right)^2 \sigma_{x1}^2 + \left(\frac{\partial f}{\partial x_2}\right)^2 \sigma_{x2}^2 + \ldots + \left(\frac{\partial f}{\partial x_n}\right)^2 \sigma_{xn}^2 \tag{7}$$

whereby it follows for the simplest and most often occuring relations that:

$$x = x_1 + x_2 \quad \text{and} \quad x = x_1 - x_2: \quad \sigma_{\bar{x}}^2 = \sigma_{x1}^2 + \sigma_{x2}^2$$
$$x = x_1 \cdot x_2 \quad \text{and} \quad x = x_1/x_2: \quad (\sigma_x/x)^2 = (\sigma_{x1}/x_1)^2 + (\sigma_{x2}/x_2)^2$$

In the case of difference measurements, e. g. for a blank correction, the total error is mostly defined by: $\sigma_x \approx \sigma_{x1} \sqrt{2}$ because $\sigma_{x1} \approx \sigma_{x2}$.

In the case of correlated parameters, the corresponding covariances have to be considered [6]. In the case of statistically determined total errors, error propagation is valid accordingly. This allows us to estimate error con-

---

[1] Here, standard deviation of the sample, in contrast to σ being the standard deviation of the population. For an estimate in the case of n measurements: $n \sigma = (n-1) s$.

tributions of single steps by the statistical partition of variances (analyis of variance [5, 6]).

5.3.3
Confidence Intervals and Uncertainty Ranges

Measurement values, following a Gaussian distribution (Eq. (4a)), may occur principally in the whole definition range $-\infty < x < +\infty$, though very important positive or negative deviations of μ show only very small probability (according to $p(x)$). Therefore, it is useful to define variation ranges including a certain number of measurement values with a given level of significance P (and therefore with a risk of error, $\alpha = 1 - P$). The statistical certainty is determined by the integration limits $\pm u(P)\,\sigma$; see for example [5, 6].

The integration regions given in Fig. 5 correspond to the following levels of significance:

$$\pm\,\sigma: \quad P = 0.683$$
$$\pm\,2\,\sigma: \quad\quad 0.955$$
$$\pm\,3\,\sigma: \quad\quad 0.997.$$

For the risks of errors α being 0.05 resp. 0.01, usually employed in analytical chemistry and quality control, the resulting limits are given by $\pm 1.96\,\sigma$ and $\pm 2.58\,\sigma$ respectively. In the case of finete sampling in analytical practice, the quantiles of the t-distribution are used as corresponding real limits (Tables e.g. [5, 6]).

The one-sided confidence interval of a mean value $\bar{x}$ is calculated according

$$\Delta\bar{x} = s_x\, t(P, f)/\sqrt{n} \tag{8}$$

with s_x being the standard deviation of a measurement series of n individual values from which $\bar{x}$ has been determined; f is the number of degrees of freedom; the level of significance P has to be determined explicitly.

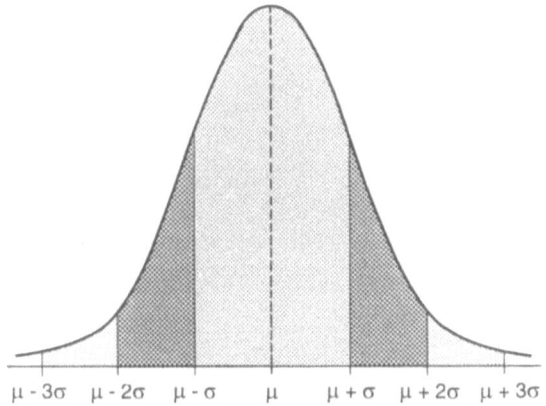

$\mu - 3\sigma \quad \mu - 2\sigma \quad \mu - \sigma \quad\quad \mu \quad\quad \mu + \sigma \quad \mu + 2\sigma \quad \mu + 3\sigma$

Fig. 5. Integration regions of the Gaussian distribution

An analytical result has always to be given with the whole confidence interval, and in the following manner:

$$\bar{x} \pm \Delta\bar{x} = \bar{x} \pm s_x \, t\,(P, f)/\sqrt{n} \tag{9}$$

Normally, this range includes $P \cdot 100\%$ of all measurement values, i.e. in the case of $n = 20$ individual measurements, one value located outside the confidence interval corresponds to the statistical expectations for $P = 0.95$.

Considering accuracy checks, e.g. with the aid of certified reference materials, the observed mean is said to be *correct* if its confidence interval includes the *true value*. If not, then the determined analytical result is *incorrect*.

Confidence intervals are only related to those analytical steps generating error contributions by repeated determinations that enter the calculation of the total error s_x. For example, if a sample is analysed n-times, the confidence interval is only related to the error of the result due to the measurement. If the confidence interval of the analytical result is intended to represent the total error of the procedure, i.e. inclusively sampling, sample preparation and measurement, then, real parallel samples have to be sampled, prepared in parallel and repeatedly measured, as illustrated schematically in Fig. 6. Then, the total error is obtained by the addition of the quadratic errors of the variations between the p samples, and, within those, by the $p \cdot q$ subsamples which are submitted to parallel sample preparation operations, e.g. separations, as well as by the variation between the total of $n = m \cdot p \cdot q$ measurements performed in parallel.

The error contributions can be determined by a simple analysis of variance (ANOVA) according to calculation schemes which, for example, are given in [5, 6].

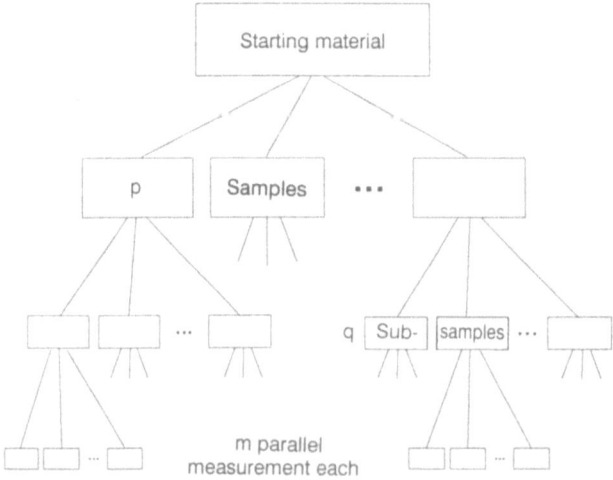

Fig. 6. Experimental scheme for the identification of error contributions

In this sense, confidence intervals are not determined in a uniform way in analytical practice; thus in the framework of quality assurance, the focus is directed onto the expression of real *uncertainties* of analytical results [7]. Generally, these can be determined by two different methods [8]:

- evaluation according to *Type A*: Determination of a *standard uncertainty* by statistical evaluation of a measurement series, expressed by the standard deviation; this evaluation is of general use, but often not sufficient,
- evaluation according *Type B*: determintion of a *standard uncertainty* on the basis of non-statistical parameters, i.e. uncertainty statements taken from certificates, literature, manufacturer's documentation or also general experiences.

In the practice of quality assurance it is necessary to determine, according to the principles of error propagation, a *combined standard uncertainty* based on experimentally (statistically) determined variance contributions (Type A) as well as on logistic error contributions (Type B, e.g. uncertainties due to volume measuring devices) [9].

5.4
Significance Tests

Statistical test procedures (statistical tests) make it possible for objective comparisons and interpretations on the basis of given experimental data. Generalizations which go beyond the given data are normally not possible.

Statistical tests are based on hypotheses, so-called *null hypotheses* H_0, the statements of which are verified by statistical tests. In this context, the following rules apply:

1. null hypotheses must always be formulated in an *affirmative* way, i.e. $H_0: \mu_1 = \mu_2$ (two subsets with the means $\bar{x}_1$ and $\bar{x}_2$ pertain to the same population $\mu_1 = \mu_2$), or, $H_0: \sigma_1^2 = \sigma_2^2$ (the variances of two subsets are identical).
2. Every null hypotheses has an *alternative hypothesis* H_A which is confirmed if the null hypothesis is rejected, i.e. if the test leads to a negative result, e.g. $H_A: \mu_1 \neq \mu_2$ (the compared mean values $\bar{x}_1$ and $\bar{x}_2$ are significantly different, thus, belonging to different populations).
3. A non rejection of a null hypotheses does *not* mean *its acceptance.* If a test does not result in a significant difference of two compared parameters, this merely means, that for reasons of the existing data, the differences are not conclusive. Evidence for correspondence is not provided. Such evidence can only be obtained indirectly, see e.g. [2, 5].
4. Each test result is only valid for a *certain* (freely chosen) level of significance P, underlying a test procedure. Thus, a test result carries a *risk* $\alpha = 1 - P$. In general, a level of significance $P = 0.95$ is chosen (corresponding to $\alpha = 0.05$). In cases where great significance or consequence is attached to the test result, a higher level of significance ($P = 0.99$) has to be chosen. The following decisions have to be made:

Table 1. Types of error for the statistical test of null hypotheses

The null hypothesis is	true	false
by the test the hypothesis is		
not rejected	test results OK	error Type 2 risk β ("concumer risk")
rejected	error Type 1 risk α ("producer risk", "false alarm")	test result OK

- H_0 for $P = 0.95$ not rejected: the difference is considered to be non conclusive;
- H_0 for $P = 0.95$ rejected: the difference is regarded as guaranteed in the normal case;
- H_0 for $P = 0.99$ rejected: the difference is highly significant.
5. Every statistical test may possibly result in two different kinds of error, as shown in Table 1, i.e.
  A) to reject the null hypothesis erroneously although it is true (*"error Type 1"*, *"false-negative"*, *risk α*),
  B) not to reject the null hypothesis by mistake, though the alternative hypothesis is true (*"error Type 2"*, *"false-positive"*, *risk β*).

Means and standard deviations calculated according to Eqs. (5) and (6) respectively, are only characteristic for sample subsets if certain pre-requisites are satisfied; the fulfillment of these can be verified by tests. The observed values of measurement series have to

a) be *normally distributed*,
b) vary at random and show *no systematic trend*, and,
c) be *free of outliers*,

this, on the one hand, in order to determine correct estimates of the mean and the variation of the measurement series, and, on the other hand, to allow the statistical comparison of these parameters to those of other measurement series.

5.4.1
Test for Measurement Series

1. *Rapid test for normal distribution*:

Range test of David

$$\hat{q}_R = R/s \tag{10}$$

with $R = x_{max} - x_{min}$ being the range and s being the standard deviation. If the calculated test value $\hat{q}_R$ is not within the limits of the tabulated values [6, 10], a normal distribution of the measurement values may not be postulated. Then, it can be tested if a normal distribution is obtained after measurement value transformation, e. g. logarithmic expression. If there is still no success, the measurement series has to be evaluated with methods of *robust statistics* [11].

The existance of a normal distribution can only be confirmed by a *goodness-of-fit test* (e.g. χ^2, Kolmogorov-Smirnov).

2. Test for trend

For a measurement series $x_1, x_2, ..., x_n$ (in the sequence of the measurement, not classified) the test parameter

$$\Delta^2 = \sum_{i=2}^{n} (x_{i-1} - x_i)^2 / (n - 1) \tag{11}$$

has to be formed. The measurement values can be regarded as independent and, therefore, as varying at random for $\Delta^2 \approx 2 s^2$; a trend has to be considered if $\Delta^2 < 2 s^2$. For an exact test Δ^2 / s^2 has to be calculated and the ratio has to be compared to the bounds given e.g. in [6].

3. Test for outliers

Among numerous tests for outliers described in the literature, in analytical practice the following have turned out to be especially useful:

a) for measurement series on a limited scale ($n \leq 25$):

test for outliers according to Dixon and Dean [5, 6],

the measurement values are sorted in ascending or descending order, depending on whether the suspected outlier value x_1^* deviates to higher or lower values. The test parameter is formed depending on the data-set

$$\hat{M} = |x_1^* - x_b| / |x_1^* - x_k| \tag{12}$$

where indices b and k may have the following values [6, 10] depending on the measurement series (n is the number measurement values):

$$b = 2 \text{ for } 3 \leq n \leq 10; \quad k = n \qquad \text{für } 3 \leq n \leq 7$$
$$b = 3 \text{ for } 11 \leq n \leq 25; \quad k = n - 1 \text{ für } 8 \leq n \leq 13$$
$$k = n - 2 \text{ für } 14 \leq n \leq 25$$

The comparison is done with the bounds $M(n; \alpha)$, e.g. in [6, 10].

b) for large-scale measurement series ($n \geq 25$):

test for outliers according to Graf and Henning:

A value x^* is rejected as outlier, if it exceeds the range $\bar{x} \pm 4 s$, where $\bar{x}$ and s have to be calculated under exclusion of x^*.

c) for nearly any measurement series ($3 \leq n \leq 150$):

 test for outliers according to Grubbs [10]

$$G = |\bar{x} - x^*|/s \qquad (13)$$

The test parameter is compared with the bounds $G(n; \alpha)$ [10].

5.4.2
Comparison of Two Standard Deviations

Two standard deviations, s_1 and s_2, with corresponding degrees of freedom, f_1 and f_2, are compared by means of the F-test. The test parameter,

$$\hat{F} = s_1^2/s_2^2, \qquad (14)$$

normally with $s_1 > s_2$, is compared to the corresponding quantile of the F-distribution [5, 6, 10]; for $\hat{F} > F(f_1; f_2; \alpha)$ it is proven that s_1 is significantly larger than s_2.

5.4.3
Comparison of Several Standard Deviations

1. for equally-scaled measurement series: Hartley test (F_{max}-test):

$$\hat{F}_{max} = s_{max}^2/s_{min}^2 \qquad (15)$$

 Tables with bounds are given in [6, 10].

2. Cochran test:

$$\hat{G}_{max} = s_{max}^2/(s_1^2 + s_2^2 + \ldots + s_k^2), \qquad (16)$$

 preferentially applied to compare k samples if one of the variances is essentially larger than all the others; Tables are given in [6, 10].

3. Bartlett test:

$$\chi^2 = 1/c \left[2{,}303 \left(f_g \lg s^2 - \sum_{i=1}^{k} f_i \lg s_i^2 \right) \right] \qquad (17)$$

with $f_g = n - k = \Sigma f_i$ being the total number of degrees of freedom (k number of groups), $s^2 = \Sigma(f_i s_i^2/f_g)$ being the weighted variance, s_i^2 being the variances of the ith group with degrees of freedom f_i and $c = 1 + \{\Sigma(1/f_i - 1/f_g)/\{3(k-1)\}\}$ [where c has the character of a correction factor; the number of degrees of freedom f_i being not too small, c approximates 1]; χ^2 will be compared to the corresponding quantile of the χ^2 distribution $\chi^2(f_g, \alpha)$ [5, 6, 10].

5.4.4
Comparison of Two Means

For the means $\bar{x}_1$ and $\bar{x}_2$ of two measurement series with n_1 resp. n_2 measurements, verification is done by the t-test. This test can only be applied without difficulty under the condition that the variances of the two sample subsets, s_1^2 and s_2^2, do not differ significantly; this has to be checked by the F-test beforehand. With the weighted average standard deviation

$$s_a = \sqrt{\frac{(n_1 - 1)\, s_1^2 + (n_2 - 1)\, s_2^2}{n_1 + n_2 - 2}} \tag{18}$$

the test parameter

$$\hat{t} = |\bar{x}_1 - \bar{x}_2|/s_a \cdot \sqrt{n_1\, n_2/(n_1 + n_2)} \tag{19}$$

has to be expressed and compared to the related quantile of the t-distribution, $t(f, \alpha)$, with $f = n_1 + n_2 - 2$.

In a similar way, an experimentally determined value $\bar{x}$ can be compared to a theoretical resp. true value μ (e.g. a reference value):

$$\hat{t} = |\bar{x}_1 - \mu|/s \cdot \sqrt{n} \tag{20}$$

In this case, the comparand $t(f, \alpha)$ has to be based on $f = n - 1$.

Equation (19) may not be applied if the variances s_1^2 and s_2^2 differ significantly. In this case, the general t-test (T_z-test) according Welch can be used [12]:

$$\hat{T}_z = \frac{|\bar{x}_1 - \bar{x}_2|}{\sqrt{s_1^2/n_1 + s_2^2/n_2}} \tag{21}$$

Again, the comparison is drawn with $t(f, \alpha)$; in this case, it is

$$f = \frac{(s_1^2/n_1 + s_2^2/n_2)^2}{(s_1^2/n_1)^2/(n_1 - 1) + (s_2^2/n_2)^2/(n_2 - 1)} \tag{22}$$

5.4.5
Comparison of Several Means

Several means $x_1, x_2, \ldots, x_k$ are compared using a simple variance analysis. The test parameter

$$\hat{F} = \frac{(n - k) \sum_{i=1}^{k} (\bar{x}_i - \bar{\bar{x}})^2\, n_i}{(k - 1) \sum_{i=1}^{k} s_i^2\, (n_i - 1)} \tag{23}$$

with k number of means to be compared,

$\bar{x}_i$ mean of the ith measurement series with n_i individual values,

s_i standard deviation of the ith measurement series,

$n = \Sigma\, n_i$ total number of all individual measurements,

$\bar{x} = 1/n\, \Sigma\, n_i\, \bar{x}_i$ weighted total mean,

as well as with $f_1 = k - 1$ and $f_2 = n - k$ degrees of freedom exceeds the corresponding quantile of the F-distribution if at least one of the means differs significantly from the others.

This global statement of variance analysis may be specified in the way to say which of the mean(s) differ(s) from the others. This is done by *multiple mean comparisons* [6].

In the simplest case, if all sample sub-groups are equally scaled, $n_1 = n_2 = \ldots = n_k$, Dixon's test for outliers is used; then, in Eq. (12), instead of the individual values, the means are entered [6]. In the case of more complicated observations with different-scaled sample subsets various procedures may be chosen; in this context see Ref. [6], page 649 ff.

5.5
Statistical Quality Assurance

In the case of product or process quality control by analytical measurements, one has to consider that not only product quality varies within certain limits which are characteristic for the production process but that analytical results also vary within the range of analytical random errors.

5.5.1
Statistical Quality Criteria

Within the scope of quality agreements, product quality is often stipulated to a standard value (target value) Q_0. If analytical control results in a lower value $x \leq Q_0$ then the quality requirement is not fulfilled and a customer may reject the product.

Owing to the variation in the results of analyses and their evaluation by means of statistical tests, however, a good product may be rejected or a defective product may be approved according to the facts outlined in Table 1 (page 114).

Therefore, producer and customer have to agree upon statistical limits which minimize *false-negative* decisions (errors Type 1, producer risk) and *false-positive* decisions (errors Type 2, customer risk) as well as test expenditure, because unlimited expenditure to obtain analysis precision and statistical security, both of which are reflected by the costs, are a burden for the producer as well as for the customer.

With the aid of the consequences resulting from wrong decisons, *risks R* can be given (e.g. in units of cost), namely, for the producer M as well as for the customer C

$$R_M = w \cdot \bar{\alpha} \cdot F_M \tag{24a}$$

$$R_C = (1 - w) \, \bar{\beta} \cdot F_C \tag{24b}$$

with w representing the priori probability that a determined quality parameter $x = Q_{C,M}$, $(1 - w)$ being the probability for $x = Q_{C,C}$; $\bar{\alpha}$ is the (onesided) producer risk (error Type 1), $\bar{\beta}$ is the (onesided) customer risk and F_M resp. F_C are parameters (e.g. costs) expressing the consequences of wrong decisions.

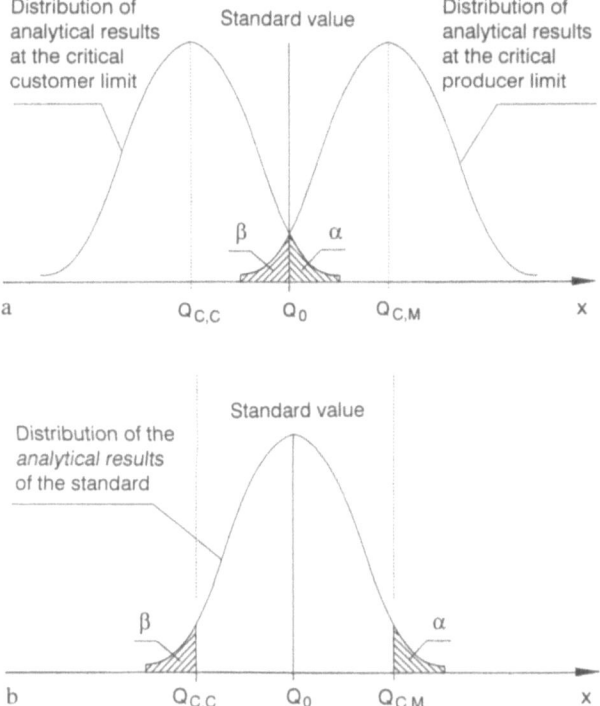

Distribution of
analytical results
at the critical
customer limit

Standard value

Distribution of
analytical results
at the critical
producer limit

β α

a $Q_{C,C}$ Q_0 $Q_{C,M}$ x

Standard value

Distribution of the
analytical results
of the standard

β α

b $Q_{C,C}$ Q_0 $Q_{C,M}$ x

Fig. 7. Distributions of analytical results for Quality Assurance

The relationship between the target value Q_0 and the distributions of the critical analyses values are illustrated in Fig. 7 for producer $Q_{C,M}$ and customer $Q_{C,C}$. For $x > Q_0$ the quality is better than agreed while $x < Q_0$ indicates poorer quality.

Because of the possibilities of error based on the risks according to Eq. (24), tolerance limits have to be fixed which, using confidence limits, are given by

$$Q_{C,C} = Q_0 - s\, t(f, \bar{\beta})/\sqrt{n} \tag{25}$$

for the lower tolerance limit with standard deviation s of the analytical procedure and n parallel determinations, as well as,

$$Q_{C,M} = Q_0 + s\, t(f, \bar{\alpha})/\sqrt{n} \tag{26}$$

for the upper tolerance limit. If customer and producer have agreed upon Q_0 not to be a target value but a *guarantee value*, $Q_{0,g}$, with fixed significance level $\bar{P} = 1 - \bar{\alpha}$, the latter has to supply at least the quality $Q_{C,M}$. The statement whether an analytical value $\bar{x} = Q_{0,g} = Q_{C,M}$ is still accepted ($\bar{x} \geq Q_{0,g}$) or not ($\bar{x} > Q_{0,g}$), is then part of the agreement, too. In these cases, the customer's riks is that in $100 \cdot \bar{\alpha}\%$ of all cases the quality can be worse than Q_0.

Quality limits are often stipulated by laws or standards, however, they are often determined by mutual agreements, too. Considering the particular risks

R_M resp. R_C, especially the values $\bar{\alpha}$ and $\bar{\beta}$ have to be consistent, whereby, in practice, $\bar{\alpha} = \bar{\beta}$ is often taken as the base.

5.5.2
Attribute Testing

In contrast to *variable testing* (comparison of measurement values), *attribute testing* means product quality testing (non-conformity tests, good-bad-test) by spot checks. Important parameters are the sample size n (the number of units within the random sample) as well as the acceptance criterium n_a, both of which are determined by the related distribution functions (hypergeometric, binomial, or Poisson distribution [10]) or by their operational characteristics in compliance with the lot size N and the number of non-confirming items p within this lot.

According to the number of defective items n_-, which have been determined in the sample it results for

$$n_- \leq n_a \quad \text{acceptance of the test items}$$
$$n_- > n_a \quad \text{rejection} \tag{27}$$

Attribute testing can also be effected on the basis of two or more (m) samples. The disadvantage for the higher expenditure to determine the sample sizes $n_1, n_2, ..., n_m$, the acceptance and rejection criteria $n_{a,1}, n_{a,2}, ..., n_{a,m}$ resp. $n_{r,1}, n_{r,2}, ..., n_{r,m}$ must be set against advantage of the possibility of coming to unambiguous decisions in clearly defined situations with only a first small random sample $i = 1$.

In contrast to classical statistical tests, three test cases exist:

$$n_{-,i} \leq n_{a,i} \qquad \text{acceptance}$$
$$n_{-,i} \geq n_{r,i} \qquad \text{rejection} \tag{28}$$
$$n_{a,i} < n_{-,i} < n_{r,i} \quad \text{inspection of another sample}$$

This model leads us to sequential tests, which generally use three exits and which represent the most rational possibility for quality control.

5.5.3
Sequential Analysis

The principle of sequential analysis consists of the fact that, when comparing two different populations A and B with pre-set probabilities for the errors Type 1 and Type 2, α and β, one examines just as many items (individual samples) as necessary for decision making. Thus the sample size n itself becomes a random variable.

Sequential investigations may be used for attribute testing as well as for quantitative measurements. The fact, to perform only as many tests or measurements as are absolutely necessary is great advantage if individual samples are either difficult to obtain or expensive, or if this is true for the measurement.

Based on the result of every individual measurement it has to be checked whether a decision may be taken or whether the studies have to be continued.

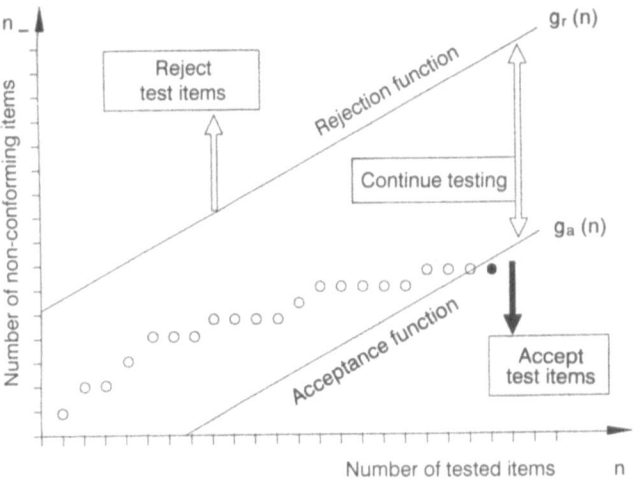

Fig. 8. Graphical evaluation of a sequential analysis

According to the final aim of sequential analysis, one differentiates between *closed sequential test plans* that always lead to a decision A > B or A < B (possibly with a large test expenditure) and *open sequential test plans*, which also allow after a certain test expenditure, the statement A = B.

Data evaluation can be done by calculation or graphically; an example of graphical sequential analysis is given in Fig. 8 for attribute testing. In the case of attribute testing, limit curves for acceptance and rejection are normally straight lines; in the case of variable testing, they are mostly nonlinear functions [10].

For variable testing, the analytical values x are represented on the ordinate, a decision is taken when the acceptance or rejection curve is exceeded.

Decisions to be taken after every individual sample are as follows:

a) for attribute testing:

$$n_{-,n} \leq n_{a,n} = g_a(n) \quad \text{acceptance}$$
$$n_{-,n} \geq n_{r,n} = g_r(n) \quad \text{rejection} \tag{29}$$
$$g_a(n) < n_{-,n} < g_r(n) \quad \text{continue testing: inspect another item}$$

with $n_{-,n}$ being the number of defective items within n (index) tested, $n_{a,n}$ resp. $n_{r,n}$ being the actual acceptance resp. rejection criterium for the current sample size n, $g_a(n, s, \bar{\alpha})$ and $g_r(n, s, \bar{\beta})$ being the acceptance resp. rejection functions.

b) for variable tests (quantitative measurements):

$$x_n \leq g_a(n, s, \bar{\alpha}) \qquad \text{acceptance}$$
$$x_n \geq g_r(n, s, \bar{\beta}) \qquad \text{rejection} \tag{30}$$
$$g_a(n, s, \bar{\alpha}) < x_n < g_r(n, s, \bar{\beta}) \quad \text{continue testing: inspect another item}$$

with x_n being the actual analysis value after n measurements, $g_a(n, s, \bar{\alpha})$ and $g_r(n, s, \bar{\beta})$ being the acceptance and rejection functions, s being the standard

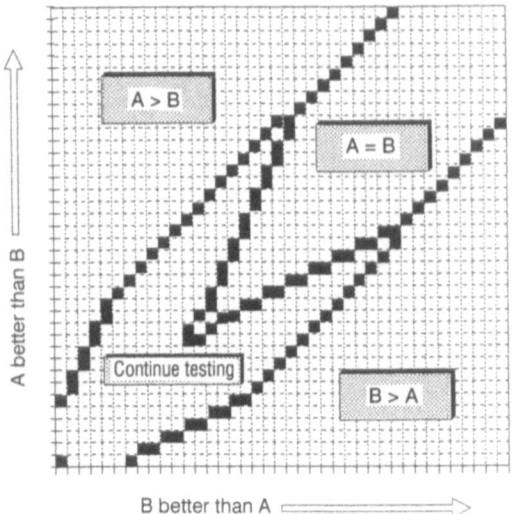

Fig. 9. Sequential test plan for direct characteristic comparison

deviation of the analytical procedure, and $\bar{\alpha}$ and $\bar{\beta}$ being the producer and customer risk, respectively.

Sequential test plan are also suitable for the direct comparison of two products or procedures based on subjective parameters as well as on measurement results, and this can be done without any calculation, but only according the following three criteria

- A is better than B,
- B is better than A,
- there is no difference between A and B.

A graphical evaluation of such comparisons is given in Fig. 9.

5.5.4
Quality Control Charts

Quality control charts were originally developed and used in industrial product control (Shewhart, 1931). Today, they serve to supervise all types of processes, production processes as well as measurement processes. Therefore, quality control charts QCC (also "control charts") are important tools for quality assurance *within analytics* itself (supervision of test methods) as well as for quality assurance of products and processes *by means of analytical chemistry*.

In each case, the interset of QCC lies in the illustrative representation of quality target values Q and their bounds. The results of sample control are depicted on the charts as a series of points; their sequence allows us to quickly discern typical situations, some of which are shown in Fig. 11.

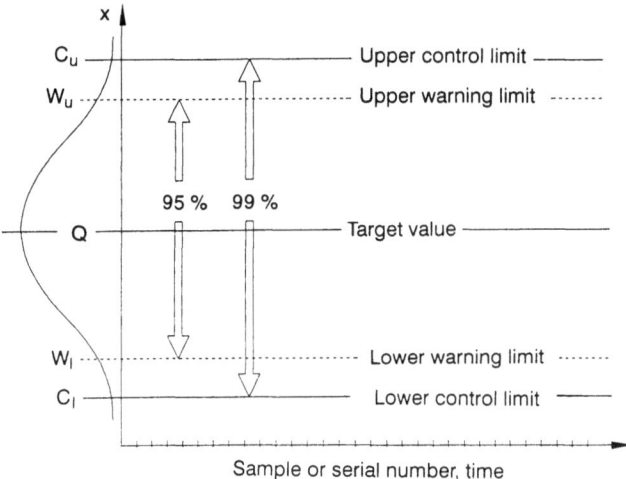

Fig. 10. General scheme of a quality control chart

QCC contain as quality target values nominal or reference values resp. optimum values as well as their limits. According to the character of the target value Q one distinguishes:

- single value charts
- mean value charts:
  - charts on $\bar{x}$
  - median charts
  - blank value charts
- variation charts:
  - standard deviation charts (charts on s resp. s_{rel})
  - range charts
- recovery rate charts
- Cusum charts
- combination charts (e.g. charts on $\bar{x}$-s and charts on $\bar{x}$-R)
- correlation charts (e.g., charts on $\bar{x}_A$-$\bar{x}_B$).

Single value charts are only used for special purposes, e.g. as *original value chart* for the determination of warning limits and control limits or, for data analysis of *time series* [14].

All other types of charts are used relatively often and have their special advantages [10, 13, 14]. The basic layout of a QCC is shown in Fig. 10; the distribution function chosen to derive warning and control limits is a Gaussian function, applicable for mean values[2].

---

[2] According to the Central Limit Theorem, mean values approximatively follow a normal distribution, even if they are obtained by non normally distributed measurement series.

Table 2. Control parameters for quality control charts

target value	limits
mean value $\bar{x}$	$L_u = \bar{x} + t(f, P) \cdot s/\sqrt{n}$
	$L_l = \bar{x} - t(f, P) \cdot s/\sqrt{n}$
standard deviation	$L_u = s_a \cdot \sqrt{[\chi^2 (n-1, 1-\alpha/2)/(n-1)]}$
$s_a = \sqrt{[\Sigma f_i \cdot s_i^2 / \Sigma f_i]}$	$L_l = s_a \cdot \sqrt{[\chi^2 (n-1, \alpha/2)/(n-1)]}$
range	$L_u = D_u \cdot \bar{R}$
$\bar{R} = \Sigma R_i / m$	$L_l = D_l \cdot \bar{R}$
$R_i = x_{i, max} - x_{i, min}$	
recovery rate	$L_u = \overline{RR} + t(f, P) \cdot s_{RR}/\sqrt{n}$
$\overline{RR} = b$	$L_l = \overline{RR} - t(f, P) \cdot s_{RR}/\sqrt{n}$
(slope of the recovery function Eq. (2))	

The corresponding factors of the t- and χ^2- distribution are reported in all text books on statistics, e.g. [5, 6, 10, 14]; for D-factors refer to [10, 13]

For the limits, the following bounds were chosen ($\bar{x} = Q$):

$$C_u = \bar{x} + t(f, P = 0.99) \, s/\sqrt{n}$$
$$W_u = \bar{x} + t(f, P = 0.95) \, s/\sqrt{n}$$
$$W_l = \bar{x} - t(f, P = 0.95) \, s/\sqrt{n}$$
$$C_l = \bar{x} - t(f, P = 0.99) \, s/\sqrt{n}$$

The limits $C = \bar{x} \pm 3 s$ and $W = \bar{x} \pm 2 s$ (according to $P = 1 - \alpha = 0.997$ resp. 0.955) are also often used.

Parameters for the most important of the above mentioned quality control charts are those given in Table 2; L_u and L_l are control limits for $P = 0.99$ and warning limits for $P = 0.95$. Figure 11 shows some typical situations for QCC. Case a represents a normal control process, cases, b, c, d, and e show situations out of control ("out-of-control-situation", OCS). In case b, a value outside the control limit demands instantaneous action. A single crossing of the warning limit only demands closer attention, while 2 out of 3 successive values exceeding the warning limit, as shown by example e, signal on OCS.

OCS also holds true if a pre-set number of successive values (7 out of 7, 10 out of 11, 12 out of 14, 16 out of 20) either lie on one side of the central line (c) or show a steadily increasing or decreasing (d) tendency.

Cases f, g and h illustrate exceptional situations, demanding instant attention as the process is threatening to get out of control. Case f shows periodic (cyclic) alterations, g a long-term trend, while in case h a particularily great number of values are close to the control limit.

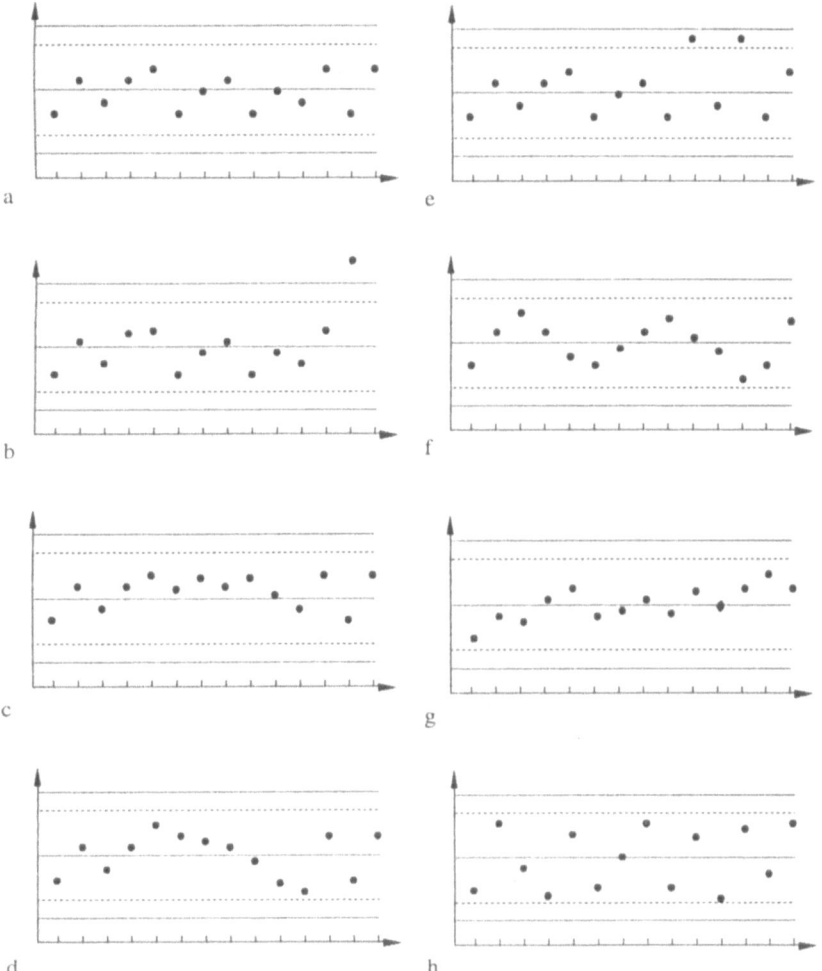

Fig. 11. Characteristic plots of control charts

Appropriately designed, Cusum control charts give a sensitive and in-structive impression of process changes. The deviations from the target value (reference value) k are cumulative; the sum for the nth sample set or series is given by:

$$S_n = \left(\sum_{i=1}^{n} x_i \right) - n k \tag{31}$$

Thus, Cusum values contain information as well on actual as on previously obtained values. Therefore, their graphical display by control charts enables to perceive earlier alterations leading to OCS than by means of the sole original values.

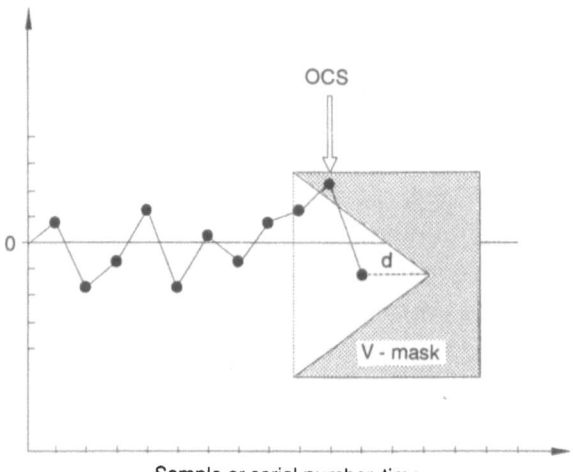

Sample or serial number, time

Fig. 12. Cusum control chart with V-mask

In connection with V-masks, OCS are directly perceived as shown in Fig. 12. The right choice of parameters (reference value, scale, V-mask angle and distance d) is a pre-requisit for effective operation [13].

Instead of the graphical evaluation, decisions based on numerical values can also be made.

5.6
Calibration of Analytical Procedures

Normally, the first step of quality assurance is calibration, i.e. the exact determination of the relation between relevant measurement parameter y and an analytical value x, expressed by the calibration function

$$y = f(x) \tag{32}$$

Assurance of accuracy is one of the central problems of calibration. It is given by the fact that equal amounts of analyte x_i in the calibration samples and in the analysis samples have to result in identical signal values y_i.

Therefore, it is usual to consider the calibration problem in a three-dimensional way [15] describing the relation between measurement values y and the "true" analytical values of calibration samples x_{true}, i.e. the *calibration function* itself, Eq. (32), as well as between determined analysis values of real samples x_{estm} and the measurement values. As a basis we used the *analysis function* which reads in its general form:

$$x = f^{-1}(y) \tag{33}$$

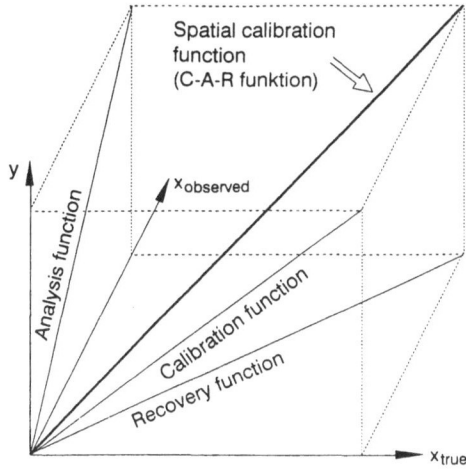

Fig. 13. Three-dimensional relationship of signals, true and observed (estimated) values x_{estm}

Furthermore, this three-dimensional consideration includes the recovery function

$$x_{estm} = f(x_{true}) \tag{34}$$

as a relation between "true" and determined analytical values, e.g. contents. Ideally, this recovery function reads $x_{estm} = x_{true}$ and represents the 45°-line through the origin. As shown in Fig. 2 and as expressed by Eqs. (1–3), in analytical practice this ideal relation may exhibit deteriorations in an additive, multiplicative and exponential way.

Calibration function, analysis function and recovery function represent the projection of the *spatial calibration function* (C-A-R-function) onto the corresponding planes (Fig. 13).

Only in the case of recovery functions $x_{estm} = x_{true}$, coincidence of the y, x_{true} – plane and y, x_{estm} – plane is observed and the calibration problem may be treated two-dimensionally, as usual.

5.6.1
Linear Fit

The relation between measurement and analysis parameters is numerically determined by means of regression. Mostly, the functional relation may be described by linear functions

$$y = a + b \cdot x \tag{35}$$

Basically, the following problems and procedures should be considered with regard to the relationship between y and x:

1) Does a relation between y and x exist and how strong is it?
  → *correlation analysis*

2) How the relation functionally be described, i.e. how can y be derived from x and vice versa?
   → *regression analysis*

Normally, in analytical chemistry, the first question is irrelevant, because a (stringent) dependency between measurement and analysis parameters, given by the measurement principle is assumed. Therefore, and because the analytical values (concentrations or contents of the calibration sample) do not represent random variables at calibration, the *correlation coefficient* r_{xy}

$$r_{xy} = Q_{xy}/\sqrt{(Q_x Q_y)} \tag{36}^3$$

which describes quantitatively the relation between two random parameters and which amounts to $-1 < r_{xy} < +1$, shows *no relevance* for analytical calibrations.

During calibration, analysis values are mostly considered to be free or almost free of errors ($s_y \gg s_x$). The regression coefficients a_x and b_x are calculated according to:

$$\hat{y} = a_x + b_x x \tag{35a}$$

where

$$b_x = Q_{xy}/Q_x \tag{37}$$
$$a_x = (\Sigma y - b_x \Sigma x)/n \tag{38}$$

The precision of the calibration is characterized by the standard error $s_{y,x}$

$$s_{y,x} = \sqrt{[\Sigma (\hat{y} - y)^2]/(n-2)}$$
$$= \sqrt{[\Sigma (y - a_x - b_x x)^2]/(n-2)} \tag{39}$$

The following further errors are important:

$$s_x = \sqrt{Q_x/(n-1)} \tag{40}$$
$$s_y = \sqrt{Q_y/(n-1)} \tag{41}$$
$$s_b = s_{y,x}/\sqrt{Q_x} \tag{42}$$
$$s_a = s_{y,x} \sqrt{1/n + \bar{x}^2/Q_x} = s_b \sqrt{\Sigma x^2/n} \tag{43}$$

Confidence intervals for a and b can be given

$$\Delta a = a \pm s_a t(f = n-2; \alpha) \tag{44}$$
$$\Delta b = b \pm s_b t(f = n-2; \alpha) \tag{45}$$

---

[3] Here and in the following regressions the following quadratic sums are used:

$$Q_x = \Sigma (x - \bar{x})^2 \qquad = \Sigma x^2 - (\Sigma x)^2/n$$
$$Q_y = \Sigma (y - \bar{y})^2 \qquad = \Sigma y^2 - (\Sigma y)^2/n$$
$$Q_{xy} = \Sigma (x - \bar{x})(y - \bar{y}) = \Sigma (x \cdot y) - \Sigma x \Sigma y/n = \Sigma (x \cdot y) - n \cdot \bar{x} \cdot \bar{y}.$$

and differences between the determined parameters and a hypothetical intercept α or slope β can be verified, where the test parameters

$$\hat{t} = |a - \alpha|/s_a \qquad (46)$$

$$\hat{t} = |b - \beta|/s_b \qquad (47)$$

are compared with the bound of the t-distribution $t(f = n - 2; \alpha)$.

While the regression parameter equals the *blank value* of the analytical
procedure $y_B = a_x$, generated either by a blind reading of the measurement
instrument or by blank contributions due to solvents, reagents or the environment, b_x constitutes sensitivity in the case of linear relationships.

In general, sensitivity S of a measurement is defined as

$$S = y' = dy/dx \qquad (48)$$

with which, for equation (35a) we obtain $S = b_x$.

Errors associated to the y-values are not constant throughout the calibration
line but depend on their distance from the calibration center $\bar{x}$. The standard
deviation of an estimated mean value $\hat{y}_x$ at position x is given by

$$s_{\bar{y}(x)c} = s_{y,x} \sqrt{1/n + (x - \bar{x})^2/Q_x} \qquad (49)$$

and constitutes the basis for the calculation of the confidenc interval (index c),
while the standard deviation for a predetermined mean value $\bar{y}_x$ resulting from
m measurements at position x is given by

$$s_{\bar{y}(x)p} = s_{y,x} \sqrt{1/m + 1/n + (x - \bar{x})^2/Q_x} \qquad (50)$$

and is used to estimate the prediction interval (index p) [12].

The confidence interval for the whole calibration line is limited by curves
according to

$$CI = \bar{y} \pm \sqrt{2\, F(f_1 = 2: f_2 = n - 2; \alpha)\, s_{\bar{y}(x)c}^2} \qquad (51)$$

The confidence range CI of a regression line is schematically shown in Fig. 14.

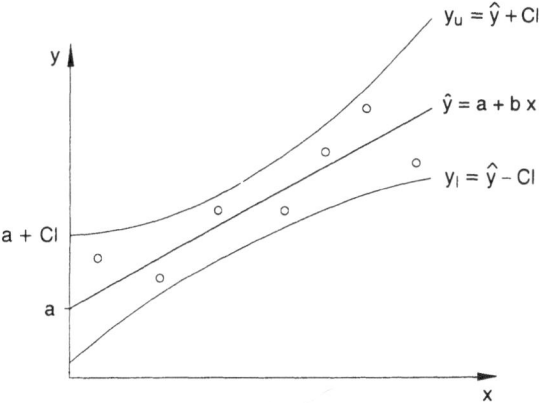

Fig. 14. Calibration linie with confidence interval CI

Linearity may be tested (after visual judgement) in various ways, namely:

a) using parameters of the linear regression: by a comparison of the variances due to the variation of the means and the variation observed within the parallel determinations [5] by means of the F-test, as well as,

b) by a comparing the variances of the residual variance of the linear model (according to Eq. (39)) with one of a nonlinear fit

$$s_{\bar{y},n}^2 = 1/f \sum [y_i - f(x_i)]^2 \tag{52}$$

with f according to the respective model (e.g. for a quadratic approach $f(x) = a + bx + cx^2$ the number of degrees of freedom $f = n - 3$) and, to be precise, either directly ($\hat{F} = s_{\bar{y},1}^2/s_{\bar{y},n}^2$), or, according to Mandel, by using the difference of the variances ($\hat{F} = [(n-2) s_{\bar{y},1}^2 - (n-3) s_{\bar{y},n}^2]/s_{\bar{y},n}^2$; compare with $\hat{F}(f_1 = 1; f_2 = n - 3; \alpha)$).

5.6.2
Limit of Decision and Limit of Detection

Various concepts and terms have been proposed to characterize the least trace concentration of an analyte that may be determined or detected in dependence on the smallest measurement value which can be distinguished from the blank value (*critical measurement value* y_{crit}) with a pre-set significance level P. In gerenal it is

$$y_{crit} = \bar{y}_B + t(f; \bar{P}) s_B/\sqrt{n} \tag{53}$$

where $\bar{y}_B$ is the mean of n blank measurements and $s_B/\sqrt{n}$ its standard deviation. H. Kaiser [18, 19] proposed $t(f; \bar{P})/\sqrt{n} = 3$, according to $P = 0.998$ for normally distributed values, and still $P \approx 0.95$ for non-normal, unimodal distributions.

Following this, corresponding to DIN 32645 [17] the limit of decision x_L is

$$x_L = 3 s_B/b \tag{54}$$

with b being the sensitivity of the corresponding calibration line. The limit of desicion according to Eq. (54) is suited as a procedural characteristic, but not suited for statistical interpretation of analytical results, for example to derive the possibly highest content, if there is no signal distinguishable from the blank. As shown in Fig. 15[4], the risk for the critical measurement value $\beta = 0.5$. In order to obtain a comparable risk of error for the Type 2 error ($\alpha = \beta$), a definition of the limit of detection x_{LD} is necessary; here, the confidence intervals of both, blank *and* measurement values (at the limit of detection) have to be considered:

$$x_{LD} = [t(f_B; \bar{P}) s_B/\sqrt{n_B} + t(f_N; \bar{P}) s_L/\sqrt{n_L}]/b \tag{55}$$

Assuming $t(f_B; \bar{P})/\sqrt{n_B} \approx t(f_L; \bar{P})/\sqrt{n_L} \approx 3$, it results

$$x_{LD} \approx 6 s_B/b \tag{56}$$

---

[4] Consider the distributions shown as being perpendicular to the x-y-plane.

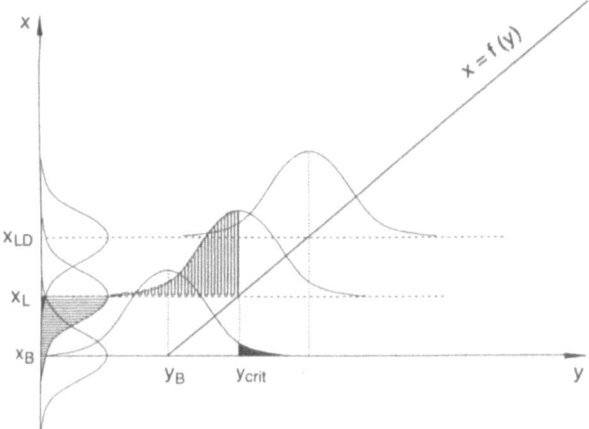

Fig. 15. Distributions of measurement and analytical values with critical limits for the definition of limit of decision, x_L, and limit of detection, x_{LD}

the so-called Kaiser *guarantee limit for purity* [18]. It has to be mentioned hat the "2 σ-detection limit", often stated by manufacturers, equals, when used as a procedural characteristic, a significance level $\bar{P} \approx 0.95$ (for normally distributed measurement values); for statistical comparisons, e.g. statements of limits, the 2 σ-limit is only based on a significance level of $\bar{P} \approx 0.68$, according to a risk $\alpha \approx 32\%$.

In the case when the limit of decision or limit of detection may not be derived by repeated measurements of the blank, their estimation, based on the confidence interval of the calibration line, is possible. While, according to equations (49, 53, 54), the limit of decision can be calculated

$$x_L = t(f = n - 2; \bar{P}) \, s_{\bar{y}(x)c}/b$$
$$= t(f = n - 2; \bar{P}) \, s_{y,x} \sqrt{1/n + (x - \bar{x})^2/Q_x}/b \qquad (57)$$

as procedure characteristic, the statistically relevant limit of detection [12] is given under consideration of Eqs (50, 55) by

$$x_{LD} = s_{y,x} \sqrt{t(f = m - 1; \bar{P})^2/m + t(f = n - 2; \bar{P})^2/n + (x - \bar{x})^2/Q_x}/b \quad (58)$$

limit of decision and limit of detection have a qualitative or semi-qualitative character of an analytical characteristic [20].

The statement of a *general* limit of quantitation, i.e. of a limit x_{LQ} calculated as a multiple of the blank variation, like $x_{LQ} = k \cdot s_B/b$, above which quantitative determinations are possible, is not very suitable, because k and, therefore, such a limit depends directly on the relevant precision, namely of a required relative standard deviation $s_{x,rel} = 1/k$.

5.6.3
Validation of Calibration Procedures

Calibration procedures have to be validated with regard to observe special requirements under which the calibration models have been developed, i.e.

basically, one has to assure by experimental studies that certain performance features (accuracy, precision, selectivity, specificy, linearity, working range, sensitivity, limit of decision, limit of detection and limit of quantitation, robustness) fulfill the requirements [21].

As *basic validation* these investigations have to be performed with the underlying calibration procedure after working out a new method to determine the reliability of the method and its possible superiority over traditional methods. In order to secure long-term stability, it is necessary to perform *re-validations*, possibly combined with the use of quality control charts, over meaningful time periods that are agred upon.

Regarding calibration, it is of special importance to characterize the following performance features:

1) *accuracy* of analysis values,
2) *precision* of calibration and analytical results,
3) *calibration model* (liner/nonlinear) and scope (working range, sensitivity, limit of decision and limit of detection).

In general, the accuracy of analytical results is assured by recovery studies. According to equation (34), for selected samples with known ("true" or "right") amounts of substance, either recovery rates or the recovery function are determined.

Precision of the calibration is characterized by the confidence interval of the estimated y-values

$$\Delta y_{x,c} = s_{\bar{y}(x)c}\, t(P;\, f = n - 2) \tag{59}$$

with the standard deviation $s_{\bar{y}(x)c}$ for a mean value $\bar{y}_{x,c}$ at position x according to Eq. (49). Analysis precision, in contrast, is expressed according to Eq. (50) by the prediction interval

$$\Delta \bar{y}_{x,p} = s_{\bar{y}(x)p}\, t(P;\, f = n - 2) \tag{60}$$

of a mean value $\Delta \bar{y}_{x,p}$ at position x (of n measurements).

The precision of analytical results is closely related to the adequacy of the calibration model. Linearity is tested by the comparing the residual variances according to criteria given in Sect. 6.1. In addition, supplementary information can be obtained by an analysis of the residuals $d_i = y_i - \hat{y}_i$. Especially by means of software packages for regression analysis, residuals can be directly visualized graphically and be assessed. If the correct approach is chosen, the residuals vary at random as shown in Fig. 16a. In contrast, Fig. 16b shows a typical case for a wrongly chosen model. Here, instead of a linear approximation, a nonlinear model should be chosen, for example.

In Fig. 16c an increase of the residual variance can be observed with increasing analytical values, a typical case of heteroscedasticity, which demands, instead of the normal unweighted calibration, the weighted regression to be chosen.

In the case of weighted regression, the dependence of the variance of the measurement values s_y^2 and the analytical parameter has to be determined in

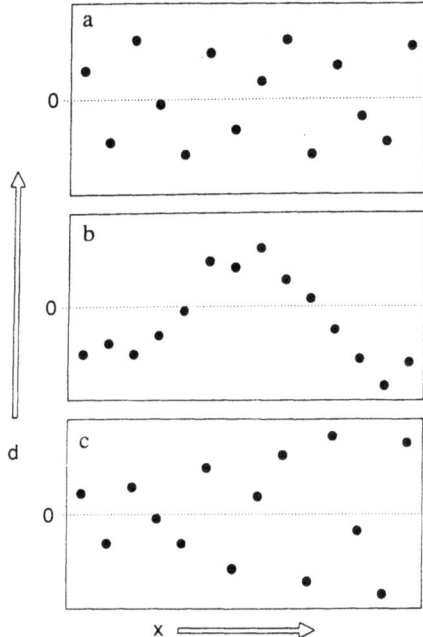

Fig. 16. Characteristic plots of residuals

preliminary studies. Then, the resulting function can be entered for weighting directly in corresponding calculation programs, e. g. in the form $w = 1/s_{\hat{y}}^2 = 1/x$.

When calculating the regression parameters by a least-square-minimization analogous to Eq. (37) to (43), one uses the variations at the individual measurement points $s_{y,i}$ to derive the weighting factors which are then introduced in the form

$$w_i = \frac{1/s_{y,i}^2}{1/n \, \Sigma \, (1/s_{y,i}^2)} \tag{61}$$

into the calculation of the weighted quadratic sums[5]. The parameters of the weighted linear regression are then obtained according to Eqs. (37) to (43) with the weighted quadratic sums, respectively. Specifically

$$a_{x,w} = (\Sigma \, w_i \, y_i - b_{x,w} \, \Sigma \, w_i \, x_i)/n \tag{62}$$

and

$$s_{y,x,w} = \sqrt{[\Sigma \, w_i (y_i - \hat{y}_i)^2]/(n-2)} \tag{63}$$

---

[5] Instead of the quadratic sums mentioned in footnote 3, page 128, the following weighted quadratic sums used:

$$Q_{x,w} = \Sigma \, w_i \, x_i^2 - (\Sigma \, w_i \, x_i)^2/n$$
$$Q_{y,w} = \Sigma \, w_i \, y_i^2 - (\Sigma \, w_i \, y_i)^2/n$$
$$Q_{xy,w} = \Sigma \, w_i \, x_i \, y_i - \Sigma \, w_i \, x_i \, \Sigma \, w_i \, y_i/n.$$

Sometimes, as an alternative to the weighted calibration, it is recommended to restrict the working range to those partial ranges with homoscedasticity (variance homogeneity). Considering the relatively high occurrence of hetero-scedastic calibration data, however, in analytical practice, the possibility of applying the weighted regression should, in general, be taken into account to a stronger extent. In this way, one obtains more reliable analytical results with a higher precision, which is constant to a large extent and covers the whole working range.

5.7
References

1. Danzer K, Than E, Molch D, Küchler L (1986) Analytik – Systematischer Überblick. Akademische Verlagsgesellschaft Geest & Portig: Leipzig/Wissenschaftliche Verlagsgesell-schaft: Stuttgart
2. Doerffel K, Eckschlager K, Henrion G (1990) Chemometrische Strategien in der Analytik. Deutscher Verlag für Grundstoffindustrie: Leipzig
3. Autorenkollektiv (1984) Analytikum. Methoden der analytischen Chemie und ihre theo-retischen Grundlagen. Deutscher Verlag für Grundstoffindustrie: Leipzig
4. Kaiser R (1971) Systematische Fehler in der Analyse. Fresenius Z Anal Chem 256:1
5. Doerffel K (1990) Statistik in der analytischen Chemie. Deutscher Verlag für Grund-stoffindustrie: Leipzig, 5. Aufl.
6. Sachs L (1992) Angewandte Statistik. Springer-Verlag: Berlin Heidelberg New York, 7. Aufl.
7. International Vocabulary of Basic and General Terms in Meterology. ISO/IEC/OIML/BIPM: Geneva, 1984
8. Guide to the Expression of Uncertainty in Measurement. ISO/TAG4/W3: Geneva, 1992
9. Wegscheider W (1993) Change of paradigms in analytical chemistry. Plenarvortrag 6th Hungaro-Italian Symposium on Spectrochemistry, Lillafüred, Ungarn
10. Graf U, Henning H-J, Stange K, Wilrich P-Th (1987) Formeln und Tabellen der ange-wandten mathematischen Statistik. Springer-Verlag. Berlin Heidelberg New York, 3. Aufl.
11. Danzer K (1989) Robuste Statistik in der analytischen Chemie, Fresenius Z Anal Chem 335:869
12. Ebel S (1993) Fehler und Vertrauensbereiche analytischer Ergebnisse. Analytiker-Taschenbuch 11 (Hrsg.: Günzler H, Borsdorf R, Danzer K, Fresenius W, Huber W, Lüder-wald I, Tölg G, Wisser H). Springer-Verlag: Berlin Heidelberg New York
13. Funk W, Dammann V, Donnevert G (1992) Qualitätssicherung in der Analytischen Chemie. VCH Verlagsgesellschaft: Weinheim, New York, Basel, Cambridge
14. Hartung J, Elpelt B, Klösener K-H (1991) Statistik. Lehr- und Handbuch der angewandten Statistik. R Oldenburg Verlag: München, Wien, 8. Aufl.
15. Danzer K (1995) Calibration – A Multidimensional Approach. Fresenius Z, Anal Chem 351:3
16. Danzer K (1990) Problems of calibration in trace, in situ-micro and surface analysis. Fresenius J Anal Chem 337:794
17. DIN 32645 Chemische Analytik – Nachweis-, Erfassungs- und Bestimmungsgrenze
18. Kaiser H, Specker H (1956) Bewertung und Vergleich von Analysenverfahren. Fresenius Z, Anal Chem 149:46

Chapter 6

Validation of Analytical Methods

Wolfhard Wegscheider

Abstract

Analytical methods, especially those applied by accredited laboratories have to be well characterized in order to clearly define their application area and the total reliability they provide. This is the primary purpose of validation.

Therefore, the most important performance characteristics, e.g. limit of detection, limits of quantitation, accuracy, precision and ruggedness have to be determined under realistic conditions, i.e. the analyst using the method in routine work, too, has to determine these characteritics in the real matrix. These data then serve to define acceptable control limits for daily measurements.

In this contribution, special importance is attributed to calibration, recovery experiments, method comparison and investigations of ruggedness.

6.1
Introduction

In practice, analytical work is used to solve many and diverse problems arising in science and technology. Accordingly, very different criteria have to be applied when assessing the individual analytical methods in order to judge their suitability in particular cases. The validation of a procedure must clarify in advance for what the procedure is suitable and for what it is unsuitable. These limitations for its applicability have to be determined as exactly as possible so that the analyst can take recourse to reliable data for the characterization of a method or clearly perceive whether, for a concrete problem, this characterization still has to be done.

Thus, validation of an analytical method helps to give us a clear definition of the modern concept of quality in order to document, in a transparent way, within the laboratory itself, as well as for the contractor, "fitness for the purpose". Validation has to be considered, on the one hand, as completion of an analytical development, on the other hand, as proof of proficiency when adopting an externally developed method; validation is therefore always necessary, regardless of whether the corresponding national and international standard procedures are applied. Also, validation is a continual process; strictly speaking, any modification of the analytical system (a different laboratory, a different piece of apparatus, a different analyst) calls for another validation. Of course, the necessary extent of (re-)validation will be tailored to the type and scope of the modifications.

In some ways, validation (and its documentation) can be regarded as being the business card of a laboratory because, apart from the formal requirements of accreditation, it clearly shows the competence of a laboratory. This is of prime importance, since from the author's own experience over a long period, the analytical instrumention does not give evidence of how qualified a laboratory is. The type, extent and conduct of a correct validation allows us to draw conclusions, not only on the existence of adequate instrumental equipment but also indicates that the analytical laboratory has a motivated management and competent employees. Especially the worth of the customer service of a laboratory becomes apparent through the extent that analytical quality objectives are oriented to the real needs of the customers who require analytical data.

6.2
Development of Analytical Procedures and Tasks of Basic Validation

Starting point of all analytical routine work is the development of analytical methods. This means an immense expenditure in time and personnel and, therefore, cannot be afforded by every laboratory. However, this developmental work is often described in related scientific publications and reports, so that a great part of this expenditure may be saved. Smaller modifications and adaptations of the equipment in one's own laboratory are mostly unavoidable and will be treated here in the same way as all other new developments. Often it is the basic principle of a procedure, determined very early in the development process, which decides on the suitability of the procedure for stable and robust application in routine work.

If, later on, the methodology of an existing procedure is to be optimized, it is very important, to identify clearly the step which is most responsible for the unsatisfactory results. Especially when considering work at trace concentration level, it is known that the early steps of analysis give rise to the most important problems (Fig. 1). Naturally, this is of significance for every laboratory because accepting a commission for sampling not only presumes the existence of the relevant competence and written sampling instructions but also the validation of this step. On the other hand, validation of an analysis of "samples – as received", e. g. without sampling and transport will start only on taking delivery.

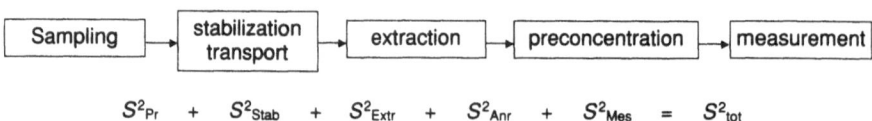

$$S^2_{Pr} \quad + \quad S^2_{Stab} \quad + \quad S^2_{Extr} \quad + \quad S^2_{Anr} \quad + \quad S^2_{Mes} \quad = \quad S^2_{tot}$$

Fig. 1. Individual steps of analysis

Laboratories working according to EN 45001 are under obligation not to accept measurement commissions when there are manifestly insufficient sampling or bad transport conditions, or no stabilization of the sample or analyte. EN 45001 makes it binding on laboratories to refuse to carry out measurements which do not lead to particularily meaningful results.

In order to present concisely the course of evaluation from a preliminary to a validated method, Taylor's graphical illustration ([1], Fig. 2) is well suited. Every method (or variation of a method) has to be classified as a preliminary one as long as the necessary validation work has not been performed.

As shown in Fig. 2, the first step is to prove experimentally that a method is "under statistical control". This will be the case if

- the mean values of the measurements are constant over a long period of time, at low as well as at high analyte concentrations,
- the precision is sufficient,
- the precision is constant, and
- all other performance characteristics (see the next paragraph) are constant, and,
- these conditions are valid not only for the measurement of calibration samples but also for the measurement of the analyte in the corresponding matrix.

Only after these requirements having been met, does it make sense to enter the next phase of validation which serves to prove sufficient accuracy. This phase usually demands more expenditure, because either other laboratories have to be called in or a completely different methodology has to be applied.

In any case, no new methodology nor a variant of an established one may be applied in analytical routine work without having completed validation and validation data disclosing its "fitness for the purpose". The necessary extent of validation is oriented on whether large or small modifications in the methodology are pending, e. g. it is based on the extent of previous validation. Though validation should be performed by the employee actually using the method in

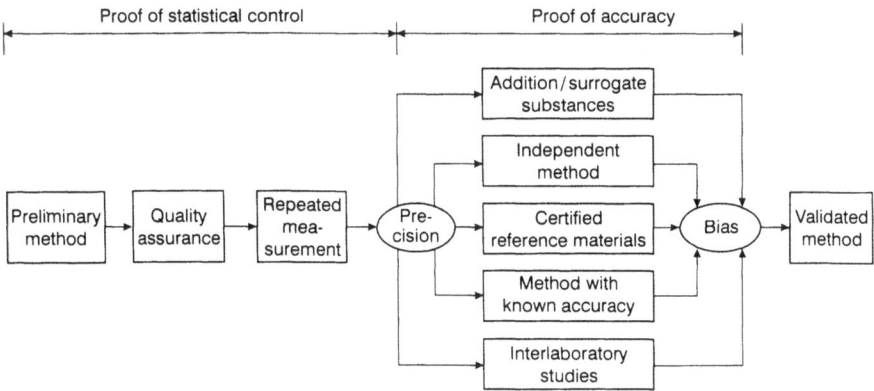

Fig. 2. Development of analytical procedures (according to [1])

practical work, after validation, the method has to be passed using the more expert knowledge of an authorized supervision of the analyst, preferably by laboratory management.

From this, as in general from the requirements of EN 45001, it follows that all procedures – and therefore all procedures modifications, too – have to be put down in writing. A clear standard operating procedure (SOP) is required to perform a validation of a procedure. This SOP should be in accordance to ISO 78/2 [2] and include, apart from instructions for experimental operation, information on the method's area of application as well as on generally known interferences, too. Hence, this standard operating procedure has to be considered as the guiding document for the scope of validation: the broader the intended application area, the more extensive validation has to be.

In summary, the task of initial validation (in the following called basic validation) is

- to prove "fitness for purpose" in the case of self-developed (or modified) methods,
- to document the analyst's competence to meet the performance characteristics prescribed by the standard in the case of well-validated standards, e.g., by recording data relating to comparability, repeatability, selectivity, precision, etc., and
- to provide in each case sufficient data necessary to set control limits for daily practice determined with regard to the performance characteristics to be met.

6.3
Validation: Definitions

Before giving full details of the validation technique of analytical methods, a short survey of current definitions should be given, because – despite certain differences – on the whole, an overall line is clearly emerging. The most important definition, which by no means constitutes the earliest or most comprehensive one, comes from the EURACHEM/WELAC document [3].

Definition of validation according to EURACHEM/WELAC [3]

---

15. VALIDATION

15.1 As well as the assessment of uncertainty for a particular method, other checks need to be considered to ensure that the performance characteristics of the method are understood and demonstrate that the method is scientifically sound under the conditions in which it is applied. These checks are collectively known as validation. Validation of a method establishes, by systematic laboratory studies, that the performance characteristics of the method meet the specifications related to the intended use of the analytical results. The performance characteristics determined include: selectivity & specificity, range, linearity, sensitivity, limit of detection, limits of quantitation, ruggedness, accuracy, precision.

These parameters should be clearly stated in the documented method so that the users can assess the suitability of the method for their particular needs. Standard methods will have been developed collaboratively by a group of experts. In theory this development should include consideration of all of the necessary aspects of validation. However, the responsibility remains firmly with the user to ensure that the validation documented in the method is sufficiently complete to meet his or her needs. Even if the validation is complete, the user will still need to verify that the documented performance can be met.

Before going closely into substance and intention of the EURACHEM/WELAC definition of validation, which also results from the further contents of clause 15 of the guideline [3], definitions originating from other fields should be mentioned. In the pharmaceutical context validation is of special significance. Here one reads [4]:

"Validation is the systematic evaluation of an analytical procedure to demonstrate that it is scientifically sound under the conditions in which it is to be applied."

or from US Pharmacopoeia [5]:

"Validation of an analytical method is the process by which it is established, by laboratory studies, that the performance characteristics of the method meet the requirements for the intended analytical applications. Performance characteristics are expressed in terms of analytical parameters. Typical parameters that should be considered in the validation of the types of assays described in this document are listed in Table 1",

The following parameter are mentioned in the quoted Table 1:

"Precision, accuracy, limit of detection, limit of quantitation, selectivity, range, linearity, ruggedness" [5].

Regarding food analysis, a Dutch provision is cited [6]:

12. Validation. The procedure which ensures that a test method is as reliable as possible. This method verification process consists of a method validation programme, which answers the question "How accurate is it?" and a method evaluation programme (interlaboratory study) which answers the question "How precise is the method in the hands of the end user?"

The new AOAC-programme, which aims at a quicker approval of modern analytical methods, mentions the following parameters to be determined at validation [7]:

Accuracy, recovery, calibration curve, linearity, limit of detection, limit of quantitation, precision, repeatability and reproducibility, sensitivity, specificity.

This comparison highlights two different aspects: a general conformity of the individual documents regarding the parameters which are important for validation, on the one hand, and, on the other hand, that the origin of most of these definitions is not primarily the German language area. Therefore, every analyst has to examine closely for his own area of responsibility whether or to what extent these guidelines are already being met. As one may learn from the above mentioned sources, validation is not only a concern of accreditation according to the EN 45 000 series in the voluntary area, but it is also so supreme importance in the regulated field (WHO, pharmacy, food, ...). This ensures, over and above the current significance of accreditation, that all areas of analytical chemistry are faced with the requirement for appropriately validated methods. This must have an effect on academic circles and scientific publishing, and this as soon as possible.

The EURACHEM/WELAC guidelines for accreditation emphasize the following: Systematic laboratory studies have to establish performance characteristics in order to allow an assessment of the meaningful application of this method. These performance characteristics comprise provisions on precision and ruggedness; all performance characteristics should disclose to the user whether the procedure is applicable to a given problem or not. Furthermore, the document takes into account that not by a long way have all standard procedures been satisfactorily validated, and, that even in the case when validation data are given, many laboratories are actually using procedure variants. In this case, again, responsibility for validation rests in principle with the analytical laboratory. Problems may arise whenever a standard method has to be applied, often because of formal reasons, despite the fact that all persons concerned – the user too – know exactly that this method is very (or completely) unsuitable. Finally, it is stipulated that also in the case of a well validated standard procedure an analytical laboratory has to furnish internal validation. In this case, the validation may be restricted to the determination of data suitable for proving the operator's competence to be adequate for carrying out the procedure.

When attempting to extract the similarities, despite all the differences occurring in the main points of the above mentioned quotations, one gets a list of quasi indispensable crucial points for all further considerations of validation. The primary contents of all documents on validation of analytical methods are therefore

- written documentation of the method,
- suitability for routine work: ruggedness of the method,
- reliability of the method, proven with real samples,
- proven command of the method within one's own laboratory (especially in the case of standard methods),
- statistical control of performance characteristics.

This list may serve as a guide for the further discussion; problem-related differences in the interpretation of the terms "documentation", "suitability", "reliability", "command" and "performance characteristics" are acceptable as long as these are oriented again on the central quality term (= fitness for purpose).

6.4
Scope and Sequence of Validation

For every analytical method applied in a laboratory, a standard operating procedure has to be available in writing [2]. This procedure is denoted as being preliminary until related validation data have proved its suitability. Generally, based on validation data, the standard operating procedure has to be audited. The points thereby affected may pertain to a modification of the procedure, but more often will relate to a specification of the application field regarding matrix and concentration range. At the same time, the results of validation serve to determine related acceptance criteria for routine analysis.

In order to establish a well-functioning analytical quality assurance, for every procedure, realistic quality objectives sustained by the results of basic validation have to be documented and pre-determined; sample control within the laboratory has to be done in such a way that analytical work is only per-formed on the basis of a concrete standard operating procedure AND of the validation pertinent to matrix, analyte and standard operating procedure.

It is well documented that an extension of the concentration range or an adaptation of the method to another matrix may easily result in incorrect measurements. Thus, if necessary, and already introduced method may need re-validation in order to assure that the extension of the application field does not affect the reliability of the method. If, from the outset, a standard operat-ing procedure is compiled in a very general way, then, the scope of validation will become greater.

Thus, the scope is subject to the number of steps within the procedure, to the type (and diversity) of matrices, and to the concentration range to be covered by a procedure. If the quality assurance system is structured in the way that a SOP only determines one single step, for example weighing, the scope of

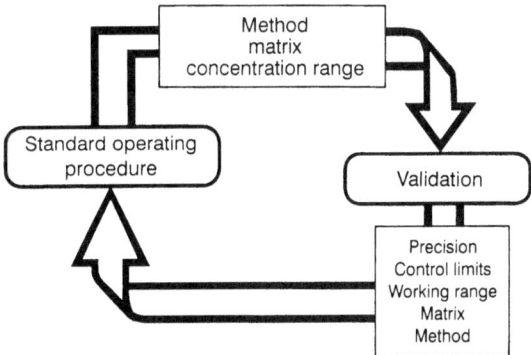

Fig. 3. Interaction – SOP/validation

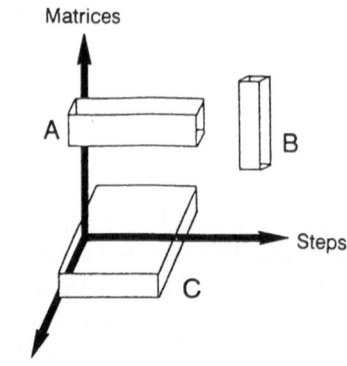

Fig. 4. Scope of validation

validation will be much smaller than it is in the case of a multi-level procedure consisting of weighing, extraction (or ashing/digestion), clean-up, preconcentration, calibration, identification and determination. An analogous consideration is true with regard to the concentration range: a method applied only to control a limit value does not need a wide dynamic range; thus a smaller amount of validation is satisfactory because only few concentration steps have to be studied. The extent of validation can be considered as the volume of a box (Fig. 4): the extension of the application area of a standard operating procedure, for example with regard to another matrix, thus entails not just additive expenditure on validation but entails expenditure which is a multiple of the original.

Typical situations in an analytical laboratory are represented by the tree cases A, B, and C. A type A procedure is restricted to a well-defined type of matrix and a relatively small concentration range, but covers many (all?) steps (control of statutory limits). A type B procedure is only related to a single operating step, but may be applied to many matrices, for example a final determination using GC or AAS with prior isoformation of the sample. Type C, finally, characterized a procedure with several steps which is valid over a large concentration range but only tested for one single matrix.

Also, when considering the sequence of basic validation steps, no stringent rules apply; but, again, some principles of analytical method development may be adopted.

These are:

- starting from one single concentration level, the whole working range should be sounded out step by step,
- starting at the final determination, one preceding step after another should be included in the validation process,
- starting with the measurement of standards, all relevant matrices should be assayed one after the other.

In every phase of this process, it has to be checked, by means of performance characteristics, if the method is still functioning satisfactorily; otherwise, further methodological improvements have to be initiated. Determining the recovery rate is a useful test for the suitability of a method. If possible, increasing amounts of analyte are added to a native sample and the recovery is determined by comparing with the additions the concentrations found when performing the analytical method. This is especially meaningful in the case of samples containing amounts of analyte lower than the detection limit.

When these operations have been completed, it only remains to prove sufficient accuracy and/or comparability. Depending on the problem, this may be done in many and diverse ways (see Fig. 2). In the simplest case, a reference material suitable with regard to matrix, analyte and concentration range is available. For small changes in the method, when one can assume that these do not affect data quality, one proves that results from the previous and new (modified) procedures are identical. This is always necessary when a well-validated standard procedure is changed internally. Further possibilities of proving accuracy consist of its comparison with a procedure different in principle or with a definitive method.

If all these ways are limited or not suitable, there only remains an inter-laboratory study for determining comparability. This presumes that all participating laboratories perform their work according to exactly identical procedures.

6.5
Performance Characteristics

Apart from the general significance of characterizing analytical methods by corresponding performance characteristics, in the framework of quality assurance these latter serve two purposes:

- on the one hand, they serve – if determined in an adequate manner – to indicate, in principle the suitability of the procedure for a certain analytical problem,
- on the other hand, they sustain related control limits and other specifications which allow one to prove that the procedure was under control for every single analysis.

Independent of whether or not these performance characteristics meet the guidelines of EURACHEM/WELAC [3] exactly, they can only be used for the above-mentioned purposes when determined in a real matrix under routine conditions. Thus, validation does not mean determining the limit of detection, limits of quantitation, precision, etc. under optimum instrumental conditions, but involves all those conditions characterizing the whole procedure, including all the preparation steps.

6.6
The Relation Between Purpose of the Procedure and Scope of Validation

Starting from these basic considerations on the scope of validation, some concrete guidelines follow which facilitate the assessment of individual cases. In order to do this, the procedures have to be classified according to their application purposes. Assuming one differentiates the following categories:

a) procedures for qualitative determination (analysis)
b) procedures for the determination of major and minor component content (assay)
c) procedures for trace analysis
d) procedures for the determination of physico-chemical parameters

requirements emarge which are grouped in the Table 1 [4]:

For definitions of the individual analytical characteristics one should refer to the compilation by Danzer [8]. Especially the definitions of accuracy, precision, linearity and limit of detection tally with those given in [3]. However, in the sense of quality assurance, selectivity is not an abstract concept, as stated in [9], but has to be proved in the course of validation by reliable and accurate measurements on real samples. If meaningful, potentially disturbing substances (interferents) are added systematically to the sample to prove sufficient selectivity; then the relationship between results and interferences is studied.

The same is true for the limit of quantitation and the working range of a procedure. Because here, one has in principle to manage without an extrapolation to high or low values, an iron law applies:

The working range does not exceed the one proven for real samples by validation.

Thus, the limit of quantitation is defined by that sample having the lowest native content of analyte for which data on accuracy and precision can be fur-

Table 1. Relationship between the purpose of a procedure and procedure characteristics to be determined

	qualitative	content determination	trace analysis	physico-chemical parameter
accuracy		×	×	×
precision		×	×	×
linearity/working range		×	×	×
selectivity	×	×	×	
limit of detection	×		×	
limit of quantitation			×	
ruggedness	×	×	×	×

nished. This defines the limit of quantitation as the lower limit of the working range. Because of the lack of reliability, purely mathematical determinations, as, for example 10 × limit of detection or 10 × standard deviation at the limit of detection, do not meet the purpose. Logically, the same applies to the upper limit of the working range, which is often extended by using dilutions. At higher concentrations, too, one cannot do without experimental data on accuracy because limitations at separation and preconcentration procedures or saturation effects during detection procedures lead to incorrect measurements.

6.7
Frequency of Validation

There are no general rules concerning either the scope of validation or its frequency. However, it is certain that one single basic validation cannot guarantee reliable long-term operation. Especially in the case of procedures not constantly run with an adequate number of samples, it is pointless to apply continuous control methods as, for example, the use of control charts.

Likewise, the multiple use of a complex instrument for more than one analytical procedure will entail all over again some new validation expenditure, especially when minor or major instrumental modifications are needed, e. g. a modification of the sample introduction systems or the detectors in chromatography. In general, special events in laboratory work will evoke a need for constant (re)validation. Without any doubt, the required scope of revalidation will depend on the degree of change in the analytical system being operated. Some of these events are listed in Table 2 and proposals are given for the corresponding validation measues.

Table 2. Measures taken in the course of revalidation in response to various events

event	measure at revalidation
new sample	internal standard, standard addition, duplicate sample
several new samples (batch)	blank(s), (re)calibration, use of a certified reference material or in-house standard material
new analyst	precision, calibration linearity, limit of detection, limit of quantitation, in-house standard samples
new instrument	check of operating specifications, precision, calibration limit of detection, limit of quantitation, in-house standard samples
new chemicals, standards	identification of critical parameters, in-house standards
new matrix	interlaboratory studies, new certified reference material, alternative methods
minor modifications of the analytical method	proof of identical results obtained with both variants (old and new) for all matrices over the whole working range

In the simplest case, a reason for validation is receiving a new sample for analysis. Depending on the method, using a single sample, one will strive for a validation by adding an internal standard, by using the standard addition method, or only by performing duplicate determinations. Per batch of samples, one (or more) blanks, a re-calibration and/or the use of a standard reference material has to be considered. When a new analyst becomes involved in a procedure special attention is needed. Here, especially for difficult sample preparation steps, the expenditure is considerable; if possible, right from the beginning, a second person should be employed who proves by the validation, too, his or her competence to run the procedure.

Depending on the type of event, there are different degrees of validation. In the case of minor changes in the analytical methodology, it is sufficient to first operate both variations over a certain period of time in parallel and to demonstrate that the values are identical. This technique is explained below in the Sect. 6.8.4.

6.8
Special Technique of Validation

6.8.1
Precision and Trueness

Precision and trueness are of paramount importance in all analytical work. However, not in all cases, are these easy to determine, and therefore, special efforts have to be made in order to obtain acceptable trueness; if, additionally, precision is also good, the method is said to have a good overall accuracy. As already illustrated in Fig. 2, several approaches exist to prove the accuracy of a method. In practice, the simplest way is to analyze a reference material. There are only two disadvantages: an appropriate reference material is not available for every problem, and, certain sample preparation steps, as well as sampling, too, cannot be controlled that way.

Another possible approach to demonstrate accuracy is the application of a totally different measurement principle. Again it holds true that only those steps can be tested which are in fact totally independent from each other. If, for example, both variations, make use of the same extraction step, then the procedures are not independent with regard to sample preparation.

6.8.2
Calibration

Considering procedures suitable for calibration, obtaining adequate calibration data is a basic principle for all validations [8]. Here, the prerequisites for a reliable calibration will be listed just once more:

a) standards (independent variable, x) are (almost) free of error,
b) the same precision over the whole working range,

c) the model is suitable: straight line or curve,
d) errors related to signals are randomly distributed,
e) errors follow a normal distribution.

The sequence of this enumeration approximately represents the importance of the criteria for analysis. If the reliability of the standard is unknown, the whole analysis based on it is of no use. The question of precision in the lower and upper working range is also very important; for example, DIN 32645 [10] may only be applied if equal precision – determined as standard deviation – is observed in the upper and lower range. This must always be tested (F-test, [8]) and, as practice shows, many methods present a worse precision (larger standard deviation) in the higher concentration range than in the lower concentration range. At a first glance this is surprising because it contradicts the intuitive conception of the analyst. He considers, however, most of the time the relative standard deviation; while the relative standard deviation decreases with increasing concentration, the absolute standard deviation measured as, for example, intensities, absorbances, concentrations or masses, show an increase. For calibration, only the latter is decisive.

In laboratory practice, one will approach the problem of different variances in two ways:

a) one decreases the working range, and – if possible – provides several narrow working ranges with separate calibration, or,
b) one derives the interdependence of standard deviation and signal intensity (and, therefore, also concentration) and uses the weighted regression to determine the regression curve.

If one favours case b) than the determination of confidence ranges has to be abandoned because they do not exist in such a case. Then of course, neither do any other parameters derived, according to DIN 32645, as, for example, decision limits and detection limits as far as they are determined by means of the confidence ranges (see also Fig. 5).

In practice it is of prime significance to examine the interdependance of precision and concentration, because, otherwise, clearly incorrect measurements will result in the lower concentration range. This is demonstrated by the following example: the determination of a PCB congener by GC-MS. At four different concentrations, standards were prepared and measured repeatedly, the lowest and highest concentration 4 times each, the two intermediate concentrations 2 times each.

This calibration, at 200 ng/ml still giving two data points which have been entered into the calculation, shows, especially in the lower region, clear differences in the plots of the curve. These differences are displayed in Fig. 5B as a less steep slope and in an intercept statistically not different from zero. This means that, according to variation A (normal regression), apparently an excess concentration is found at lower concentrations.

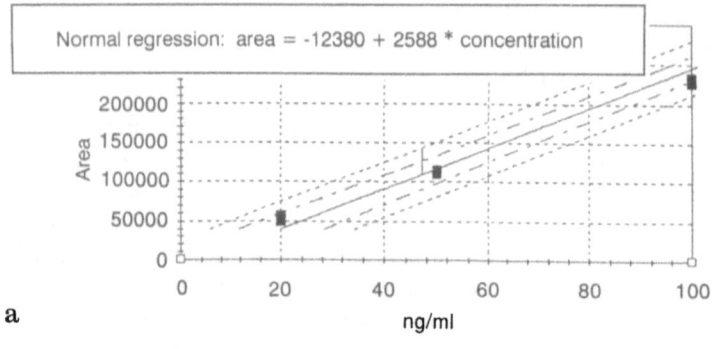

a

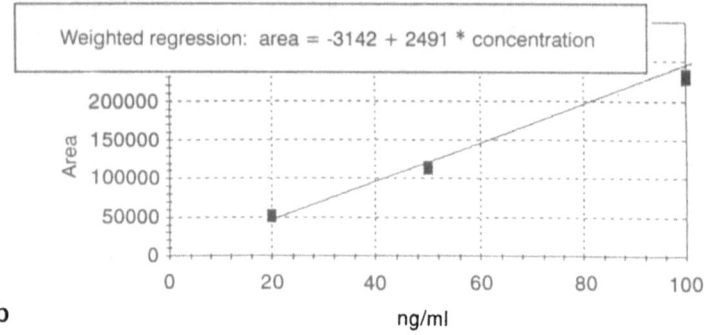

b

Fig. 5. Normal, A, and weighted regression, B. Weighting using the inverse variance at each concentration value. Data for 200 ng/ml are not shown for scaling reasons

6.8.3
Recovery Studies

Recovery studies are another very important means of documenting the relia-bility of a method. Increasing amounts of analyte are added to aliquots of a sample with a very low content of this analyte, each aliquot is mixed thorough-ly. Subsequently, each sample aliquot is submitted to the procedure and the recovered amount is compared to the added amount. Principally, the following boundary conditions have to be kept in mind:

1) the analyte has to be added in the same chemical form as (probably) the native form present,
2) it must be possible to realize a good homogenization of the sample,
3) the native content – if possible – has to be below the detection limit.

If it is not possible to fulfill the conditions in point 1 and point 2 for the whole procedure, it may be meaningful to carry out at least those steps of the proce-dure for which these prerequisites can be fulfilled. If, for example, the addition is done after a dissolution or extraction step, then the subsequent analytical procedure steps could be still validated.

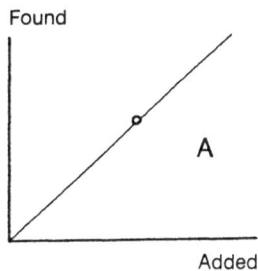
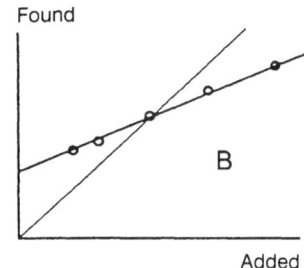

Fig. 6. Significance of recovery rates A = one concentration; B = several concentrations

Table 3. Data for recovery determination

x_recov µg/L	y_recov µg/L	recovery rate %	recovery µg/L
0	– 1.508166611		– 1.423464586
0	– 2.002261757		– 1.423464586
2.5	0.962309117	38.49236467	1.152812775
2.5	0.303515589	12.14062358	1.152812775
5	4.585673518	91.71347035	3.729090136
5	3.432784845	68.65569689	3.729090136
10	9.361926592	93.61926592	8.881644859
10	9.19722821	91.9722821	8.881644859
20	19.73792465	98.68962324	19.1867543
20	19.73792465	98.68962324	19.1867543
40	38.84293694	97.10734236	39.79697319
40	39.99582562	99.98956404	39.79697319

For recovery studies it is important to perform several additions with different amounts; this prevents one ascertaining a satisfactory recovery which is really false (Fig. 6).

This experimental series results in a recovery value for each concentration of the working range which at least – if sufficiently validated – can also be used to correct the final results. The following example gives such an evaluation for the determination of Al in Ca-gluconate. The procedure is based on graphite furnace atomic absorption measurement and is intended to demonstrate which deviations have to be accounted for when using a direct calibration against a Triton solution. The raw data are listed in Table 3, the numerical data evaluation is shown in Table 4.

Table 3 lists the concentrations and signals for calibration, and, in the next two columns, the concentrations and signals for the matrix (free of analyte). These data are now evaluated in an appropriate way: a regression determines the relation of added and found analyte; then, this information is compared to the target values. If the results are in accordance with the target values, then the curve must have a slope equal to 1.0 (within the statistical variations) and

Table 4. Statistical evaluation of recoveries

---

calibration function 1ˢᵗ degree (y = a + bx)

slope		1.030510944 µg/L(µg/L)
CI (Slope)	1.001587726	1.059434163 µg/L/(µg/L)
(intercept)		− 1.423464586 µg/L
CI (intercept)	− 1.968579661	− 0.878349511 µg/L
mean value (x)		12.91666667 µg/L
mean value (y)		11.88730178 µg/L
residual standard deviation		0.617161098 µg/L
standard deviation of procedure		0.598888446 µg/L
relative standard deviation of procedure		4.636555711 %
t-value (95 %)		2.228139238
Qx		2260.416667 µg/L^2

control of analytical precision

standard deviation (cal)	0.50842065
F-comparison	1.473501975
crit, F-value (99 %)	4.849141533
no significant difference 99 % level	

test for systematic deviations
WARNING: constant systematic deviation
WARNING: proportional systematic deviation

---

the intercept at the origin. This can be checked by a statistical test. The results are listed in Table 4.

Table 4 comprises all relevant results of recovery studies and has been taken from [11]. Rounding off the significant digit may be omitted because, on the one hand, these are intermediate results, and, on the other hand, the significance is explicitly supported by statistics.

This evaluation reveals an insufficient recovery especially in the lower concentration range. This is signalled for the user in the last two lines by a "warning". Then nomenclature used is according to the standard; hence, here it is only referred to [10, 12]. In this case, too, as always in data evaluation, a graphical presentation is very meaningful.

It is obvious from Fig. 7 that precision is quite good while recovery is insufficient in the lower concentration range.

Thus, a direct calibration without taking into account the matrix may not be used; an application of the standard addition method may possibly be successful.

6.8.4
Comparison of Methods

While totally new procedures may not be introduced into a laboratory without an extensive validation, for less important changes and modifications, the question arises, whether these will influence the data at all and, thus, accuracy and precision. Such changes are especially related to deviations from standard

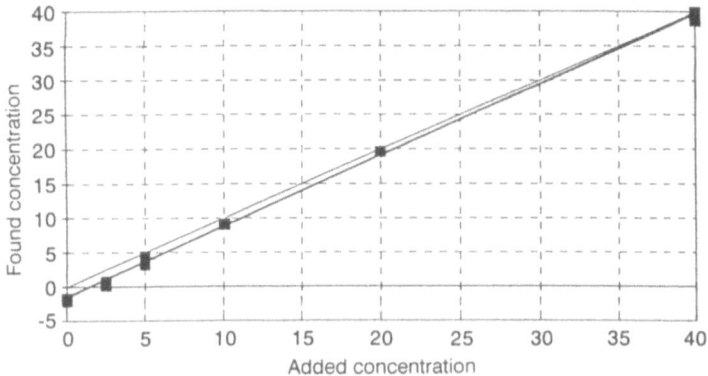

Fig. 7. Plot of the recovery function
upper curve   = ideal recovery
lower curve passing through the measured values = real recovery

procedures. A possibility of checking this is to perform both, the new and old variant in parallel until there is sufficient evidence that no differences emerge or that the new variant not worse than the one currently applied.

A simple check is the t-test [8, 13]. Here, for each sample (k) submitted to both procedures for reasons of comparison, the difference is calculated between the results obtained by the original method (subscript i) and its variation (j). For sample k, thus. $\Delta x_{ijk} = x_{ik} - x_{jk}$ and subsequently, the mean and the standard deviation of all the differences are calculated. Calculating subsequently the confidence interval [8] for the differences, the value 0.0 has to be included. If not, a difference is proven.

The precision of both methods, too, has to be checked by the F-test [8]. Otherwise, it could be that the (apparent) correspondence is accompanied by a marked deterioration of precision; this fact, too, would indicate that both procedures (procedure variations) do NOT give comparable data.

After having studied these two preliminary assessment criteria, the data set is checked for occurring proportional and constant deviations by a special regression with the assumption that errors are associated to the results of both procedures. This may be checked by applying either the method of orthogonal regression [14, 15] or a robust regression according to Refs. [16–18]. Again, both methods are explained in Ref. [13]. By no means, may classical regression be used for method comparison, because this leads to useless results [14].

Regression results are used to check if an intersection differing significantly from zero occurs on one axis. This would lead us to conclude the existence of a *constant error* between both procedures. The next thing to do is to check whether the slope equals unity or not. Deviations from this slope of ideal correspondence (1.0) are called *proportional errors*.

As an example here are mentioned two procedure variations for the determination of lead in blood using graphite furnace atomic absorption spectrometry.

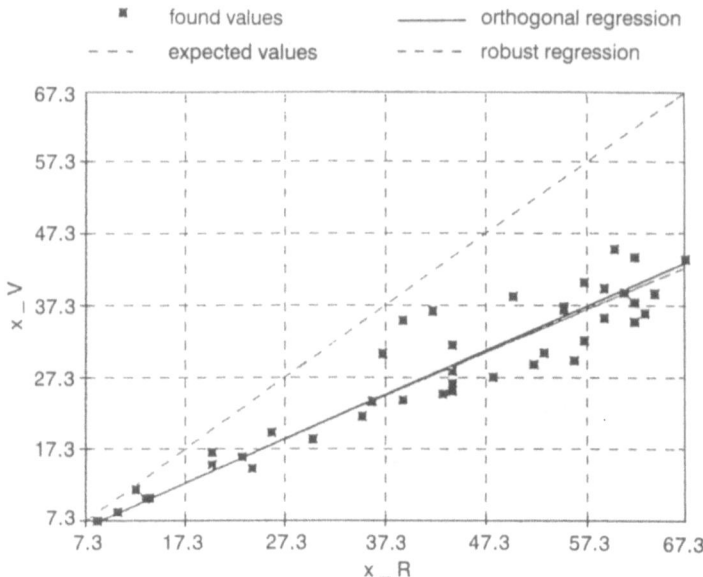

Fig. 8. Comparison of methods – Lead in blood

As illustrated by Fig. 8, in this case, orthogonal regression and robust regression lead to very similar results. In the presence of outliers, i.e. samples which show an excessively low or high result using one of both methods, preference should be given to robust regression.

Both evaluations indicate – as shown by Table 5 – that the procedure variations are NOT identical. The mean values are different and there are proportional systematic deviations recognizable by the slopes and their confidence intervals:

Orthogonal regression:
Slope 0.6159; CI (slope): 0.550, 0.682; intercept 1.90
Robust regression:
Slope 0.6025; CI (slope): 0.536, 0.68; intercept 2.227; CI (intercept); – 0.69, 4.424.

Thus, in this case, inconsistency of the procedures is obvious and there is no need for further statistical back-up. In general, care has only to be taken that these comparison studies are not based on too few samples. A minimum of 12 – 24 samples should be respected because, otherwise, the danger of not being able to recognize the differences will be too high.

6.8.5
Ruggedness

Ruggedness of a method is a criterion which is gaining more and more importance for routine analysis. In short: ruggedness means that the quality of

Table 5. Results of statistical evaluation

Orthogonal regression			Robust regression		
mean value (x_R)		42.9833333	slope		0.602537
mean value (x_V)		28.3761905	CI (b_rob)	0.536	0.68
standard deviation (x_R)		17.2962412	intercept		2.2269556
standard deviation (x_V)		10.6537709	CI (a_rob)	– 0.69	4.424
difference (mean)		– 14.607143	Cusum test		
slope		0.61595874	Cusum (i, max)		4
intercept		1,90023077	critical value (95%)		8.81380735
standard diviation (difference)		8.14036593	linear dependence		
rel. random error (%)		18.938424			
rel. constant systematic error (%)		– 33.983271			
rel. proportional error (%)		38.4041263			
correlation coefficient		0.93991664			
residual variances		2.64309752			
crit. t-value (95% N-2 d. f.)		2.02107458			
variance of the slope		0.03244857			
CI (slope)	0.55037775	0.68153973			
WARNING: proportional systematic deviation					
t-value (comparison of the means)		– 11.629097			
crit. t-value (95% N-1, d. f.)		2.01954208			
WARNING: significant difference in the means					

data is independent of small operating variations of the procedure. As shown by Table 1, page 144, ruggedness is essential for every procedure, independent of whether it is applied for qualitative or quantitative analysis.

Ruggedness may be determined by one of two different ways:

1) In an interlaboratory study with a sufficiently large number ($\geq$ 8) of participating laboratories following one and the same procedure, random variations will occur always in their operation.
2) In one's own laboratory, running a carefully planned series of experiments by varying the most important experimental parameters within the predetermined (or possibly occurring) limits of tolerance and, then, by studying the effects on the results.

Of course, case 1) involves higher expenditure since more laboratories have to be involved. This will not always be possible, and, therefore, some time ago, AOAC proposed a methodology to be used in case 2) [13] which is very efficient because it is operable with a minimum of assays. This methodology is also found in very modern guidelines [7]. Hence, it will be discussed in detail here.

The basic idea consists of the fact that due to efficient experimental planning, only approximately the same number of experiments have to be performed as there are potential factors. In practice, this means performing 8 experiments for 7 factors, 12 experiments for 8 – 11 factors, 16 experiments for 12 – 15 factors, and so on. If certain non-linear effects are also to be studied, the number of experiments is increased accordingly. The ruggedness study proceeds as follows:

1) define those variables (factors) which most probably influence the result;
2) for each of these variables define, the maximum tolerances to be accounted for in routine work,
3) draw up an adequate experimental plan (according to [19]),
4) perform and evaluate experiments in order to determine the influential factors,
5) take measures either to keep exactly within the target values or for further optimization of the procedure with the aim to reduce the influence of the corresponding variables.

Here, an example of liquid chromatography is given: the determination of a pharmaceutically important substance after its separation from some impurities; additional interest is attached to the quantification of these impurities.

The HPLC separation is performed by a gradient elution at pH 7.0 in acetonitrile. In the first two columns of Table 6 are listed the parameters of interest and their normal values ("regular conditions").

The maximum tolerances for which a stable operation of the procedure is still expected are shown in the following two columns. It can be seen that there is a total of eight factors. Thus a minimum of 12 assays would have been sufficient. Here, a slightly broader approach with 16 assays is used. These 16 assays are listed in Table 7 giving evidence for a completely balanced experiment with exactly 8 assays at the upper limit (+ 1) and 8 assays at the lower limit (– 1) for each factor. This, naturally, results in very good statistics when comparing the two extreme conditions. Thus, the first assay is performed using the "– 1" conditions for all parameters: 0.0% MeCN initial concentration, 18% MeCN final concentration, buffer 0.05 mol/l, pH 6.8, flow rate 1.3 ml/min, 30 °C, detection wavelength at 225 nm, and an injection volume of 5 μl.

The balanced character of the experimental plan is further demonstrated by the fact that all effects can be estimated independently. This characteristic

Table 6. Chromatographic conditions

Parameter	regular conditions	lower limit – 1	upper limit + 1
mobile phase: acetonitrile [%]	from 1 to 20	0 to 18	2 to 22
buffer capacity [mol/l]	0.1	0.05	0.15
pH value	7.0	6.8	7.2
flow rate [ml/min]	1.5	1.3	1.7
temperature [°C]	35	30	40
detection at [nm]	230	225	235
injection volume [μl]	5	5	15

Table 7. Study plan for 8 factors and 16 assays

No.	Factor								resolution
	A	B	C	D	E	F	G	H	
1	− 1	− 1	− 1	− 1	− 1	− 1	− 1	− 1	4.6
2	1	− 1	− 1	− 1	− 1	1	1	1	9.5
3	− 1	1	− 1	− 1	1	− 1	1	1	4.7
4	1	1	− 1	− 1	1	1	− 1	− 1	7.5
5	− 1	− 1	1	− 1	1	1	1	− 1	5.6
6	1	− 1	1	− 1	1	− 1	− 1	1	9.7
7	− 1	1	1	− 1	− 1	1	− 1	1	1.1
8	1	1	1	− 1	− 1	− 1	1	− 1	8.5
9	− 1	− 1	− 1	1	1	1	− 1	1	5.3
10	1	− 1	− 1	1	1	− 1	1	− 1	9.5
11	− 1	1	− 1	1	− 1	1	1	− 1	2.1
12	1	1	− 1	1	− 1	− 1	− 1	1	7.1
13	− 1	− 1	1	1	− 1	− 1	1	1	5.4
14	1	− 1	1	1	− 1	1	− 1	− 1	7.7
15	− 1	1	1	1	1	− 1	− 1	− 1	1.3
16	1	1	1	1	1	1	1	1	6.7

is not obvious because it is based on the orthogonality of the study plan. Therefore, it is not advisable to draw up such a study plan by one self, but one should take recourse to approved software solutions [19]. This has the further advantage of a possible computer-aided data evaluation. Now, laboratory organization is challenged to perform the experiments exactly according to this plan and to determine the most important characteristics of the system under these various conditions. This chromatographic determination was concerned with the resolution of the individual components. The chromatogram under normal conditions is shown in Fig. 9.

Assuming a very critical separation between "impurity 3" and "impurity 4", then the result of the ruggedness studies will indicate factors influencing this separation. The resolution of these two peaks is shown in the last column of Table 7.

The evaluation is now carried out by determining the (mean) difference of resolution for '+' level and '−' level experiments for each factor. The effect of factor A (initial concentration of acetonitrile) results from:

$$(- 4.6 + 9.5 - 4.7 + 7.5 - 5.6 + 9.7 - 1.1 + 8.5 - 5.3 + 9.5$$
$$- 2.1 + 7.1 - 5.4 + 7.7 - 1.3 + 6.7)/8 = - 0.47.$$

The effect of factor B (final concentration of acetonitrile) is calculated according to:

$$(- 4.6 - 9.5 + 4.7 + 7.5 - 5.6 - 9.7 + 1.1 + 8.5 - 5.3 - 9.5$$
$$+ 2.1 + 7.1 - 5.4 - 7.7 + 1.3 + 6.7)/8 = -0.70.$$

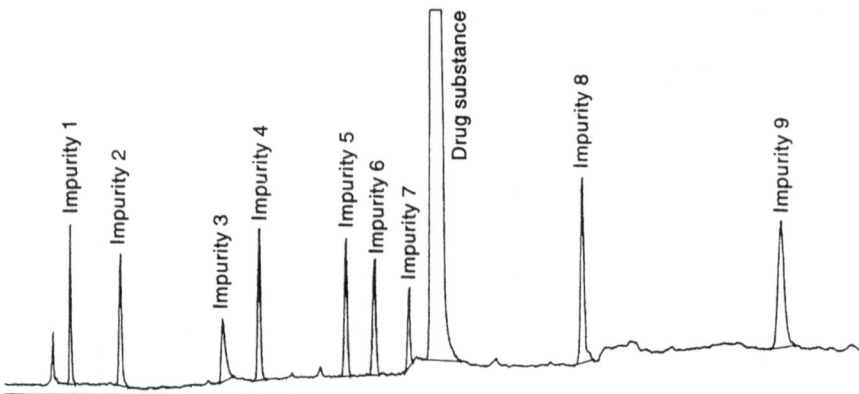

Fig. 9. Chromatogram under normal conditions

The total result of this series shows all influences on the resolution of "impurity 3" and "impurity 4":

molarity	=	4.57791	± 0.380886	pH	= − 2.22209	± 0.380886
$MeCN_{initial}$	=	− 0.472091	± 0.380886	$MeCN_{final}$	= − 0.697091	± 0.380886
flow rate	=	0.602909	± 0.380886	temperature	= − 0.597091	± 0.380886
injection vol.	=	1.04655	± 0.399243	wave length	= 0.402909	± 0.380886

These figures state that resolution either increases (positive values) or decreases (negative values) with an increase of the corresponding factor from the lower tolerance limit (– 1) to the upper tolerance limit. The ± values are standard deviations. This evaluation shows directly that chromatographic resolution is very critical with regard to buffer molarity and pH because these both factors have the most effect. Probably, in this case, a satisfactory resolution can only be realized when strictly observing the target values for these two parameters. However, for other factors (e. g. gradient, temperature, flow rate, wavelength) the tolerances are absolutely acceptable.

Of course, in practice still follows an evaluation of the influences on the resolution of other pairs of substances. Because this is done analogously, no details will be given. No further experiments are needed because all characteristics of a separation can be observed simultaneously. Therefore a ruggedness study demands no special expenditure: 16 additional experiments were sufficient to obtain all the information related to ruggedness.

However, if the results reveal that the method is not sufficiently robust for practice, then additional method optimization has to be performed with a view to its practical suitability. This is still less expensive than accepting failures during production (routine application of a method).

6.9
Conclusions

No method can be applied by a laboratory without the laboratory having performed a pertinent validation. The elements of validation, shown in this chapter, have to assure that with adherence to relevant regulations of quality assurance, the laboratory, with every application of the method, produces data which are well-defined with respect to precision and accuracy. This presumes that every laboratory not only validates its own procedures but that it can also sufficiently prove by its own data its own competence with regard to the application of standard procedures. In particular, this requires regular re-validation which should be performed in a flexible way according to the specific circumstances. Special attention has to be drawn to the results of (apparently small) procedure variations in order to assure that the results obtained either by the variants or by the standard procedure do not deviate from each other. On the other hand, here again, the precision of the procedure variation is the most important criterion for its acceptance. A laboratory accredited according to EN 45001 has always to be in a position to prove the mentioned criteria by documentation.

For providing the mentioned data, I am indebted to the following institutions: ASA-Arbeitsgemeinschaft für Spurenelemente, Graz; Glaxo Analytical Evaluation, Ware UK; Working Group on Chromatography, Institute for Analytical Chemistry, Microchemistry and Radiochemistry of the TU Graz (M. Wenzl). I thank Ch. Rohrer for his cooperation during the development of the software package on validation.

This chapter is based in part on work on quality assurance of the research group "Edelmetallemissionen" of BMFT (Bonn), Förderkennzeichen 07VPTQS, supported by GSF, Munich.

6.10
References

1. Taylor JK (1983) Anal Chem 55:500A
2. ISO 78/2, Layout for standards – Part 2. Standards for chemical analysis, Genf
3. EURACHEM/WELAC Guide 1, Accreditation of Chemical Laboratories, Laboratory of the Government Chemist, London 1993
4. The Validation of Analytical Procedures Used in the Examination of Pharmaceutical Materials, WHO/PHARM/89.541/Rev. 2, Genf 1989
5. US Pharmacopoeia, USP XXII, <1225> Validation of Compendial Methods, USP-Commission
6. Validation of Methods, Inspectorate for Health Protection, Food Inspection Service, Niederlande, September 1992
7. AOAC Peer-Verified Methods, Policies and Procedures, AOAC International, Arlington 1993
8. Danzer K, chap 5
9. Otto M, Wegscheider W (1986) Anal Chim Acta 180:455
10. DIN 32645 Nachweis-, Erfassungs- und Bestimmungsgrenze, Mai 1994

11. Handbuch ValiData, EXCEL-Makro zur Methodenvalidierung, Graz 1993, c/o ASA-Arbeitsgemeinschaft für Spurenanalyse, Schörgelgasse 53, A-8010 Graz, Austria
12. DIN 38402 Teil 51 Kalibrierung von Analysenverfahren
13. Wernimont GT, Spendley W (1985) Use of Statistics to Develop and Evaluate Analytical Methods, AOAC, Washington
14. Cornbleet PJ, Gochman N (1979) Clin Chem 25:432
15. Haeckel R (1981) Das Medizinische Laboratorium 14:8
16. Eisenwieder HG, Bablok W, Bardoff W, Bender R, Markowitz D, Passing H, Spaethe R, Völkert E (1983) Laboratoriumsmedizin 7:272
17. Eisenwieder HG, Bablok W, Bardoff W, Bender R, Markowitz D, Passing H, Spaethe R, Völkert E (1984) Laboratoriumsmedizin 8:232
18. Bablok W, Passing H, Bender R, Schneider B (1988) J Clin Chem Clin Biochem 26:783
19. e.g. Statgraphics, manugistics Inc., 2115 East Jefferson Street, Rockville, MD 20852, USA

Traceability of Measurements to SI: How Does It Lead to Traceability of Quantitative Chemical Measurements?

Paul De Bièvre

Preface

The drastic increase in attention for the concept of "traceability" and its introduction into the large world of chemical measurements has been one of the more remarkable features of recent years. The concept of traceability in physical measurements has been found useful for many years, but its exact meaning has been open to varying interpretations. In chemical measurements, however, it has not been applied.

In this study an attempt is made to explain what traceability means – or could mean – in chemical measurements. It is believed that quantitative chemical measurements (in analogy with – quantitative – physical measurements) should be made traceable to the SI unit of amount of substance, when the entity measured can (with adequate precision) be identified by chemical formula and that automatically means measurements of amount of substance, the relevant basic quantity in our international measurement system (SI). The problem of traceability of other chemical measurements must be treated separately.

In this paper and in all amount-of-substance measurements, traceability to the mole is implied and should be required even when the uncertainty of that bond (to SI) is larger than for the measurement link in the field. The mole is as constant as the kilogram (to a relative uncertainty of about 10^{-7}) and provides a virtually timeless anchor for all measurements. The mole is operationally approachable for all amounts of all substances and therefore provides an ideal world-wide measurement reference. Who would want to send samples of an ozone RM around the world for uniform comparisons and suppose that it would stay constant for a decade? Looking at many important effects in nature leads us straight back to the fact that counting of entities is superior to description by mass. Consider for instance:

1. the many alloys that depend for their properties on certain fixed valency electron/atom ratios;
2. the huge number of isomorphous oxides, sulfides and tellurides (including the silicates, the stability of which compounds is largely determined by the ratios of cations that fit into 4-, 6- and 8-fold coordinated positions;
3. the number of hydrophilic groups per unit area that will determine surface tension, the number of condensed benzene rings that may affect toxicity, the number of monomers per average polymer chain that determines fibre tensile strength etc.;

4. all free-radical reactions, a function of the number of unpaired electrons; and
5. all radioactive processes for which the number of entities leads to straight-forward description of events connected with decay.

These examples show that the entity named may not necessarily be elements, ions, or specific compounds. It this not an important attraction of measurements by the mole?

In some cases a deep-anchored traceability may not be needed: a local or sectorial one may be sufficient. When a concept is logically built up in this study up to and including traceability to SI, it is because a comprehensive work model – which includes possibilities for various kinds of traceabilities – may be useful for international discussions on comparability across borders and over long periods of time. The study is also drafted out of service-mindedness to the analyst who wants to be sure that there is a basically sound international structure of chemical measurements in which he operates, so that he can concentrate on his own main and daily problems where there are far greater uncertainty sources to struggle with.

7.1
Introduction

If we want to communicate with each other about knowledge and pass it on to future generations, it is mandatory to describe the objects and phenomena we observe, both *qualitatively* and *quantitatively*. Stating that object A is "larger" than object B is proof of a deeper insight than just talking about objects A and B being "different". If, in addition we can say: the length of object A is 5 times larger than the length of object B, we deepen our knowledge still further. We start *quantifying* by using multiplicating (or dividing) factors (in this case: 5). Note that we have in fact been determining a *ratio* between two lengths and that did not require the use of a unit! If we now use a scale along which we measure the length of object A by using the length of object B as a reference, we use the length of object B in order to "measure" the length of object A and we could in fact do so for any other object. It is a convenient and orderly way of putting the lengths of many objects on the same scale, because we then express lengths as multiples of a common unit. Stating a length requires multiplication of the determined ratio with a unit. The determination of a ratio followed by the multiplication of that ratio with a unit is a *measurement*.

The length of object B then becomes the "unit" by convention (a human agreement) and usually gets a name. Measuring ratios of the same quantity e.g. length in the same unit meter is a good form of metrology enabling us to attain small uncertainties. It is also very convenient. The fact that we can now compare many different objects (for their length in our example), is called *comparability*. This comparability has become possible because there is a *trace* linking the length of any object to the length of object B (our unit). The existence of this generates the concept of *traceability*. It is easy to see that traceability is a condition for direct comparability: if the length of object A were traceable to, say, the ell (used for centuries in the linen trade) and the length of object B to the foot, then the lengths of objects A and B would not be directly comparable since they have not been measured in the same units. Appropriate conversion factors can remedy such a situation of course: if the ell and the foot are traceable to a common unit, or to each other, traceability is re-established. The full "trace" in this case is constituted by two links and hence is a little longer: from the linen to the ell and from the ell to the common unit. The "trace" now runs along a chain of links. Since each link has its own uncertainty when measured, the total uncertainty of the trace is equal to the total uncertainty of the chain, i.e. of the combination of the uncertainties of all links. *Traceability and uncertainty are two concepts which are inherently coupled in any traceability system.* If both measurements discussed above are not traceable to the same unit, it could give rise to injustice – and hence problems – in trade and commerce. It would be very uneconomical to manufacture bolts under one system of units and make them fit the nuts made under another system. The use of the same unit permits the direct comparison of scientific observations, or *scientific measurements*. Thus it is convenient that everybody uses one and the same, unalterable, unit. It makes measurement results independent over space and – equally if not more important – over time when the unit is well chosen for its constancy.

Note that the measurement process of comparing two lengths (or masses) to each other, can also be carried out by looking at differences. One accumulates a count of length (or mass) units for each of the objects, then measures the remaining difference. This can actually be a superior process (smaller uncertainty).

It was – and is – convenient to have an officially agreed list of units (metre, second, Kelvin, ...) to measure the quantities (length, time, temperature, ...) we want to measure. It is also convenient to use it consistently! We have

Table 1. The quantities and units of measurements. By convention physical quantities are organized in a dimensional system built upon seven *base quantities*, each of which is regarded as having its own dimension. These base quantities and the symbols used to denote them are as follows:

physical quantity	symbol for quantity	name of SI unit	symbol for SI unit
length	l	metre	m
mass	m	kilogram	kg
time	t	second	s
electric current	I	ampere	A
thermodynamic temperature	T	kelvin	K
amount of substance	n	mole	mol
luminous intensity	I_v	candela	cd

managed to reduce the number of quantities which we observe in nature to seven and we made these into an official list. All other quantities such as speed, acceleration, volume, density, etc. are a combination of these seven base quantities. This list is the heart of the SI system. It is given in Table 1 and was decided by the Conférence Générale des Poids et Mesures (10th CGPM, 1954, 11th CGPM, 1960 and 14th CGPM, 1971). The quantity[1] with which we are concerned in "chemical measurements" is *amount of substance*, its unit is the mole (symbol: mol). It was fixed as such by the 14th CGPM in 1971. Note that it is only applicable to a specified entity (e.g. a well specified element or compound).

7.2
Traceability of Chemical Measurements: the Problems

The scope of this study is to deal with traceability of quantitative chemical measurements i.e. measurements of amount of substance. There are other measurements categorized under "chemical" e.g., the calibration of a mass scale in a mass spectrum or chromatogram, or a wavelength in an optical spectrometer. These have to do with identifying a given species from a complex mixture (a chemical material) in order to do a quantitative measurement on it. Only quantitative chemical measurements come under the SI system, i.e. when an uncertainty can be assigned. However, one can very validly argue that "identifying the species of interest (the "entity") is as much a part of the concept of mole as is counting" (G. Price, personal comment).

A difficulty in introducing "traceability of chemical measurements" is that chemists are used to describing their amounts of substance (amounts of chemical compounds) in terms of weight or mass [1].

The reason for the use of weights was – and still is – that the balance provided such an extraordinarily simple tool for comparing "weights" of substance through "weighing", with very small uncertainty. Weighing, however, ignores the chemical nature of the measurand and the measurement of "amount". Since it is a fundamental consequence of the particulate nature of matter that (numbers of) particles interact with each other in chemical reactions and not masses of matter[2], ratios of numbers of reacting particles are simple – and – important. The ratios of weights (or masses) of these reacting particles are not. Determining ratios of quantities expressed in the same unit can (still) be done to a much higher accuracy than determining ratios of numbers of particles by direct counting. In order to be able to use this simplicity of ratios of numbers of particles, the old chemists devised the concept of atomic (and molecular) weight. It allowed them

---

[1] The term "quantity" in the meaning "amount" is used extensively in English and French. This causes confusion since quantity in the Englisch language also has another meaning (see Table 1). In French and in German, this other meaning is expressed by a different word: "grandeur" and "Grösse".

[2] not all chemical analyses do necessarily trace back to chemical reactions.

a) to carry on using the balance to weigh amounts
b) to convert the ratios of weights to ratios of numbers which were identical to the ratios of the numbers of the reacting particles on the atomic scale.

The further logical conclusion from the discovery of the particulate nature of matter *had to be* that a unit for amount of substance was a number of particles or entities,

- either   *1 (one entity* (unit: 1): a good choice in principle and correspon-
            ding to the reality on the atomic scale, but useless on the macros-
            copic scale where we handle much larger amounts of substance; we
            cannot (yet) count individual particles (atoms, molecules, ions, ...)
            conveniently enough in order to be useful for macroscopic analyt-
            ical chemistry; that it is a very convenient concept for the mind as
            a way of describing reality, is illustrated by the following example:
            it may be agreed by consensus not to calculate in a socio-eco-
            nomical study the averate earning per habitant, but per kg or ton
            of population [2]; this somewhat less satisfactory approach - in
            fact very impractical - when using the results of the study, im-
            mediately springs to mind;
- or   *{N_A} entities* where N_A is the Avogadro Constant,
            $(6 \cdot 022\,136\,5 \cdot 10^{23}\,mol^{-1})$: a good choice for handling "weighable
            amounts" on the macroscopic, visible scale and chosen such that it
            would indicate the number of particles in a "weight" of so many g
            as the atomic or molecular weight indicates (gram rather than kg
            was convenient); such an amount was called one mole (symbol:
            mol).

Note that any fraction of N_A would also be suitable as unit. It is, however, convenient to select the number of particles present in 12 g of ^{12}C rather than in 6 g of ^{12}C or 3 g of ^{12}C.

Thus, we are facing a practical - de facto - situation where a "wrong" instrument to measure amount, namely the chemical balance, is

a) very convenient to determine ratios through an *approximation,* which consists of measuring ratios of "weight" then converting them into ratios of amount by means of "atomic weights" or "molecular weights", and
b) accurate enough for most chemical purposes: weighing *accuracy* and knowledge of atomic weights (see Fig. 1) easily provide an uncertainty of 10^{-3} which is more than sufficient in most cases

The possible inaccuracy of the approximation is the inaccuracy of the conversion factor weight $\rightarrow$ amount (Fig. 1) which is negligible as long as natural isotopic compositions are involved. Putting this in equations:

$$\frac{N_1(E_1)}{N_2(E_2)} = \frac{n_1(E_1)}{n_2(E_2)} \tag{1}$$

we see that the ratio of numbers $N(E_i)$ of particles of elements E_i (or compounds) is the same as the ratio of moles $n(E_i)$. This ratio can also be found

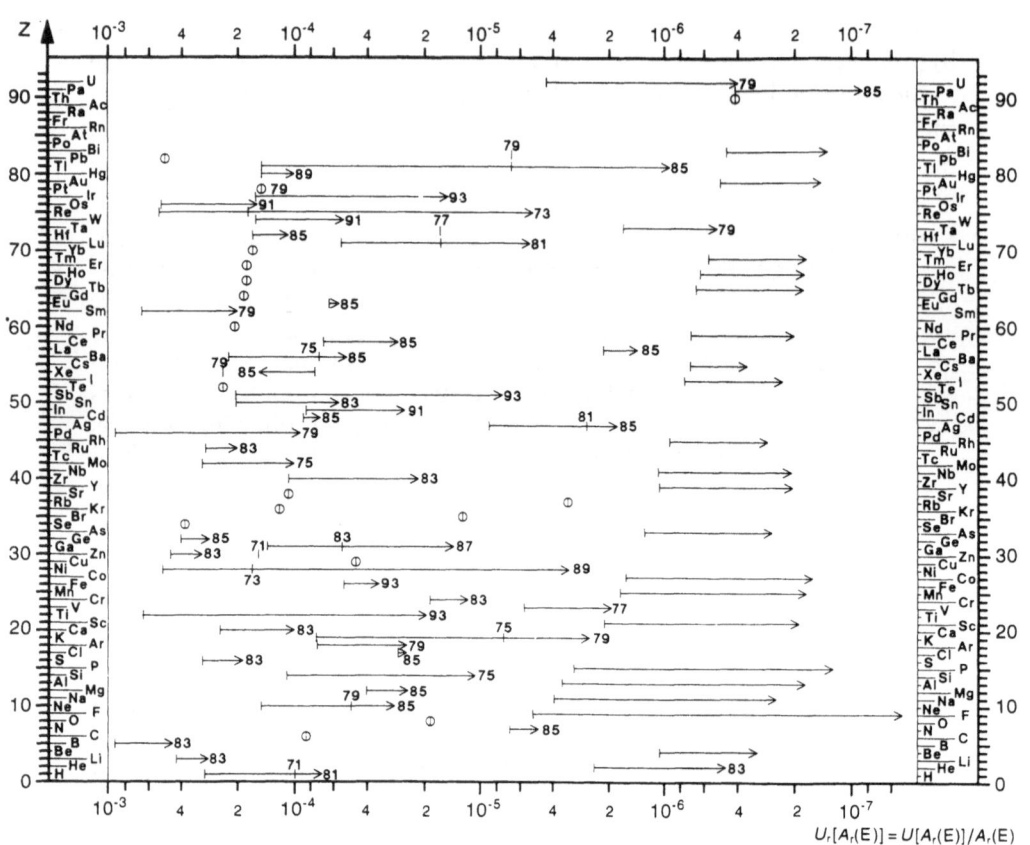

Fig. 1. Changes in relative uncertainties, $U_r[A_r(E)] = U[A_r(E)]/A_r(E)$, of the IUPAC recommended atomic weights of the elements from 1969 to 1993.
$U_r[A_r(E)] = U[A_r(E)]/A_r(E)$ is the abscissa, plotted logarithmically; the atomic number, Z, the ordinate. The element symbols are indicated in the left and right margins. ① indicates no change. The length of bar equals the $U[A_r(E)]/A_r(E)$ improvement factor (deterioration only for Li and Xe) from 1969 to 1993. The last two digits of the year of the last change are always indicated to the right of the bar (except for the monoisotopic elements). Intermediate changes for all but the monoisotopic elements are indicated on the ordinate by short vertical lines. The years in which these intermediate changes were made are indicated in the figure above or below these lines. Although 65 elements were given more precise standard atomic weight values since 1969, the uncertainties of 23 elements remain in excess of 0.01 %

through the "approximation", i.e. through the ratio of "weights" or "masses" $m(E_i)$:

$$\frac{m_1(E_1)/M(E_1)}{m_2(E_2)/M(E_2)} = \frac{n_1(E_1)}{n_2(E_2)} \qquad (2)$$

where $M(E_1)/(M(E_2)$ is the conversion factor (ratio of molar masses M) needed in the approximation.

Using weights as a surrogate for "amounts" is, conceptually, an approximation. In addition, it must consider the uncertanty of the conversion factor (typically $\leq 10^{-3}$) which in most cases is insignificant. It s important to have a quantitative idea of the uncertainty of this conversion factor: in Fig. 1 the present uncertainty of our Atomic Weights (Molar Masses) is given for all elements. Uncertainties of Molecular Weights can easily by computed from this Table.

Thus, in practical work, weighings are convenient and fit for the purpose. When we are however, structuring our basic measurement system for traceability, we obviously search for the "trace" which leads in the shortest possible and most transparent way from the daily, routine measurement to the relevant SI unit for the quantity we are interested in without "approximations". In a metrological sense, weighings are a very practical approximation of the most direct way, but always an approximation, in fact, an indirect way. They may only be fit for purpose when the "direct way" is unfit. Note also that the approximation becomes less valid when the isotopic composition of the chemical compounds – and hence their molar masses (atomic weights) – change, a phenomenon which has been discovered on a small scale in nature and has been carried out on a larger scale by men on certain elements. Quoting analysis results in mass units is somewhat of a " "barbarisme métrologique" [2].

There is a second phenomenon which should be noted when using "weight" or "mass" in basic metrology of chemical reactions, and that is that in any chemical reaction, the masses of the reacting compounds change since there is energy uptake or production. Again, the resulting effect is (extremely!) small, but it would hurt basic metrological and scientific thinking to use *changing* masses in order to designate *constant* amounts. Also on this ground, using mass rather than amount, is an approximation.

There is a third and more basic objection to the use of mass (and hence of the kg) in chemical metrology or analytical chemistry. It is the fact that mass is a property of matter which has to do with inertia. A number of particles (amount) has nothing to do with inertia. Amount of material was prior to inertial mass, a recent (17th century) conceptual invention. Although it may be argued that "substance" in the quantity "amount of substance" has inertia, this is not what was incorporated in either the concept of the definition of "amount of substance" or of the unit "mole". The word substance was added because of the necessity to state to what "amount" is referring. When, in an analysis, we are after a number of particles – an amount – we should not express our results in a unit which is of a mechanical nature [2]. To examine and describe nature, we have decided, by the adoption of the SI system, to scientifically and formally (and legally!) identify "mass" and "amount of substance" as two basically different quantities with their own basic units "kg" and "mol". Everybody, including chemists, ought to stick to this system unless and until we have

a) very good reasons to change that (something which would obviously require a rather big change in our insight into nature)
b) made a proposal to change to the relevant authority (CGPM)
c) decided to implement this change by international agreement.

A fourth problem arises from the distinction between "chemical" and "physical" measurement, or, rather, from what has been made into a distinction. Looking at the basic quantities of our SI system (Table 1), there is no trace of such a distinction. Why should there be any? At most, one could say that chemical metrology is a subset of metrology that measures (potential) chemical reactants. As we have seen in the introduction, measurements have to do with quantifying qualitative observations. Chemical measurements or analytical chemistry are quantifying in the same way as "physical" measurements do. Consequently, since we are used to length measurements, etc., we should also get used to amount measurements.

7.3
Physical and Chemical Measurements: Is There a Difference in Principle?

It appears more and more that "chemical" measurements have been termed so because they were carried out by chemists and "physical" measurements by physicists. So maybe we are looking more at a so-called distinction between "chemists" and "physicists" than at basic differences in the kind of measurements we perform. This makes this distinction appear all the more artificial in scientific terms. However, it is true to say that in chemical metrology, chemical properties are critical in any consideration of the measurement process.

   What could have been possible reasons for such a distinction?

– Is "amount of substance" a different concept in physics and chemistry? Clearly no: a number of entities is neither especially "physical" nor "chemical"; if anything it is mathematical. When it is attached to a unit, it becomes part of a measurement result.
– Does "amount of substance" cover such operations as chemical separations? Measurements are by definition *quantifying* something and separations do not quantify. However, it is important to point out that chemical separations i.e. separating different entities or species *before* measuring (= quantifying) them, is a very important step on the way of quantification of these species (e.g. molecules). For reliable and accurate measurements, a 100% complete separation is, in many cases, even an essential condition for accurate quantification, but in itself it is not a measuring process. However, the uncertainty of the separation process, i.e. the possible incompleteness of the separa-

tion, must be included in the uncertainty budget of the measurement. Since an uncertainty can be assigned to this step, we should consider it as part of the measurement process.

The preceding consideration leads to what *may appear* as a rather important distinction between chemical and physical measurements: measuring an amount of an identified species is mostly a measurement of the amount of a single species in a mixture of species (called a matrix). It is a measurement of one, single component of a complex system which is under examination. In the practice of a chemical measurement, *the whole complex system is the system under examination*. This is why the chemical *separation* process – separating the one single component from the complex system – is so important: it simplifies the amount-of-substance measurement of the single component. Is that not where chemistry started altogether? The Dutch word for chemistry is "Scheikunde" which comes from the word "scheiden" = to separate. "Scheikunde" is the ability to separate. In fact, quantitative (= analytical) chemistry was only born *after* separation processes were developed and this became an art, then a science, pursued by chemists. Time has now come where amount-of-substance measurements join the measurements of other basic quantities already going on in physics. Large areas of chemistry do not currently fall in this category: molecular structure elucidation, toxicological effects, designation of a crystallographic type etc. It is important to distinguish quantitative chemical measurements (amount measurements) from these other chemical measurements. It is to be expected, though, that what is now called "qualitative" may be semi-quantified, then fully quantified in the future, either through better measurement (counting?) methods, and/or the conception of new units.

7.4
Traceability of Measurements: Are There Precedents?

Having identified some of the problems with the traceability of chemical measurements, let us now examine some of the fields where traceability has long been established and even formally organized.

In mass measurements, all mass determinations must be traceable by international convention (and by law!) to the prototype of the kilogram, the definition of which is: "the kilogram is the unit of mass; it is equal to the mass of the international prototype of the kilogram" (3rd CGPM, 1901). This prototype is kept at BIPM (Bureau International des Poids et Mesures) at the Pavillon de Breteuil in Sèvres (near Paris). The system was organized such that, in a number of countries, national standard institutes have a "national kilogram" which is traceable (Fig. 2) to the "international kilogram" at Sèvres by way of a regular comparison (every 30 years). Each of these national institutes is responsible for acting as a traceability centre for its country: all mass measurements have to be traceable to the "national kilogram" by law. Note that the comparability of mass measurements between field laboratories in Fig. 2.

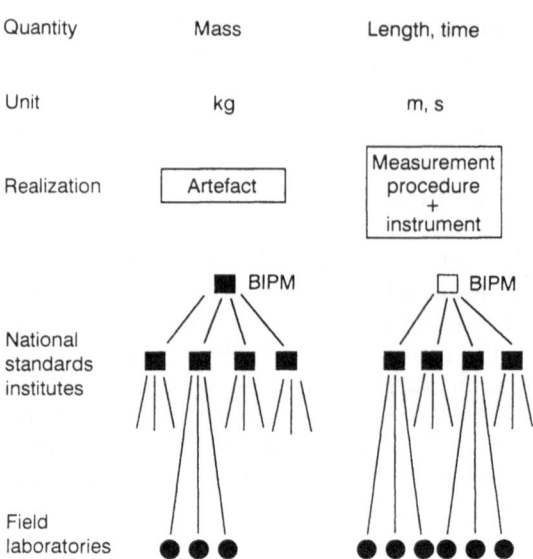

Fig. 2. Traceability of mass, length and time measurements

is realised by the fact that all their measurements are traceable to the SI unit by being linked to the *"realization" of the SI unit* which is the prototype of the kilogram in Sèvres. Note that the realization of this SI unit is an artefact: a man-made piece of metal.

In length measurements, at the instruction of the "Convention Internationale du Mètre" (1875), the 1st CPGM 1889 established the metre as unit, defined to be 1/40 000 000 of the circumference of the earth represented as the distance between two scratches on a Pt-Ir bar preserved at the BIPM (Bureau International des Poids et Mesures) at the Pavillon de Breteuil in Sèvres (near Paris). By this international convention – which is implemented in the law of a number of countries – all length measurements had to be traceable to the metre, more exactly to its *realization* (the distance between two scratches on the Pt/Ir bar). The system was further organized such that the national standards institutes of those countries had a "national metre" which was traceable to the international metre at Sèvres by way of a regular comparison (Fig. 2). Each of these national institutes was – and is – responsible to act as traceability centre for its country: all length measurements must be traceable to the "national metre". Since the distance between the two scratches soon became too imprecise for modern measurements, a new definition adopted in 1960, (11th CGPM) redefined the metre as the distance across an exact number of wavelengths of an energy transition in ^{86}Kr. In 1983, the 17th CGPM redefined it again, now as "the length of path travelled by light in vacuum during a time interval of 1/299 792 458 of a second", thereby linking length to time. These new definitions will gradually change our basic traceability concepts and requirements (because "realization" of the unit can now be achieved in several laboratories).

In a similar way, all time measurements, actually time interval measurements, must be traceable to the second, originally defined as 1/86400 of the mean solar day, later of the mean tropical solar day, then as "the duration of 9192631770 periods of the radiation corresponding to the transition between the two hyperfine levels of the ground state of the ^{133}Cs atom" (13th CGPM, 1967). Traceability of time interval measurements means the linking of all such measurements to the time unit second. Again BIPM and (some) national standard institutes act as key points of a network achieving traceability by linking measurements in the field to the second. Traceability to the unit second requires a realization of the definition of the unit, which is a little more difficult than in the case of the kg or the (former) metre, but oddly enough much more like the case of the mole.

We note that the SI unit metre is no longer linked to a man-made artefact – the Pt/Ir metre – but to nature-made velocity of light. The SI unit for length is now anchored in an unalterable, indestructible characteristic of nature, i.e. in a fundamental constant (the velocity of light on earth in vacuum). A similar thing has happened to the second: it too is anchored in an indestructible characteristic of nature. What is now the "realization" of these theoretical definitions of SI units, since linking actual measurements in the field to the units needs practical "realizations" of these units? Two visible signs on a pillar on the Place de la Concorde in Paris (1792) guaranteed that everybody had access to a new universal unit for length measurements, the metre: a simple and effective form of "realization" of the unit! As already noted, the recent evolution is one where definitions anchor the base SI units in characteristics of nature (such as fundamental constants). Realization of a unit in the form of an artefact was simple, visible and made it readily available as realization of the unit. Traceability to this realization was easily comprehended. Anchoring definitions of units in fundamental characteristics of nature, requires a re-thinking of realization. One sees that the realization becomes a *measurement procedure on an instrument* (count of a defined number of transitions in ^{133}Cs, count of a defined number of wavelengths in ^{86}Kr, later a defined velocity of light). We note the trend in realization of the units in areas where traceability has been a long standing reality: from artefacts to measurement procedures on an instrument. Even for the kg – the last artefact of our SI system, getting away from the artefact is already under discussion: a definition of the kg as a number of ^{12}C atoms is likely in the not too distant future. It will be possible to do this when the number, $\{N_A\}$, has been determined to $\leq 5 \times 10^{-8} N_A$. Again a (described) measurement procedure on an instrument will be needed and this will constitute the realization of traceability to the unit.

In summary:

1. there are precedents for the traceability of measurements;
2. traceability is needed to ensure that all measurements of the same quantity are expressed in the same units; it is obvious that this is a condition for comparability;
3. BIPM has a key role in the realization of the SI units and traceability to them;

4. the tendency in realization of the SI units is from "man-made" (artefacts) to "nature-anchored" (fundamental constants);
5. measurement procedures implemented on a specified measurement instrument play a role of fundamental importance in the realization of SI units and hence in the establishment of traceability.

Attention is drawn to point 5 above, which will play a very important role in amount ("chemical") measurements.

7.5
Traceability of Amount Measurements: Present Status

Although amount of substance is the internationally agreed quantity in our SI system (Table 1), the term "chemical measurements" continues to be used almost universally. It would be more consistent with our international agreement on measurements (SI) to systematically use "amount (-of-substance) measurements" instead, similar to "length measurements", "time measurements", "current measurements". The consistent use of the correct term "amount-of-substance" (reminder: of a specified "entity"!) in analytical chemistry would contribute considerably to the clarity in describing analytical work. An expression such as "amount concentration" leads naturally to, for example, "$mol \cdot kg^{-1}$". It indicates an amount per mass (of chemical matrix). Some say that we cannot go over to a more consistent use of the mole for an amount concentrations since that would imply that all other components of a matrix would have to be expressed in mol too. Consequently, the question arises: should we describe the matrix also as an amount or can we keep "mass" to qualify the matrix? There are at least four good reasons to keep "mass" for the matrix:

a) mostly it is one component in a matrix which we need to determine because of its action as a chemical agent, not the matrix itself;
b) most of the chemical matrices are so complex that it is impossible to talk about amount: all species in the matrix would have to be qualitatively and quantitatively known in order to do this; that time is still far off and, besides, it is not needed;
c) the practical simplicity and wide availability of balances in the daily practice make a mass determination easy, cheap and sufficiently accurate: it is simply convenient, it is "fit for the purpose";
d) When very "primary" CRMs have to be characterized, weighing is the start of the path to amount, but before converting mass to amount for the main component (the "entity") concerned, corrections for impurities on the total mass can conveniently be done on a weight basis: impurities can be treated as weight corrections to more than sufficient accuracy (no change from mass to amount needed for them).

From the earlier sections, it has gradually become clear that, in terms of an internationally organized system for traceability of amount measurements, there is nothing in existence for chemical measurements which comes even close to the organization of traceability of measurements of the six other SI quantities. What chemists have been doing, is using reference materials (RMs) or certified reference materials (CRMs) to correct their quantitative measurements for systematic errors, thus making their measurements "traceable" to these RMs. However, this approach is not always reliable because in many cases the RM does not have the "same" matrix as the unknown sample. In addition, it then makes us question the traceability of these RMs to our SI system, a problem which has not been addressed convincingly for many CRMs. Inter-

Fig. 3. Measurements of amount (of substance) not comparable: where is the traceability?

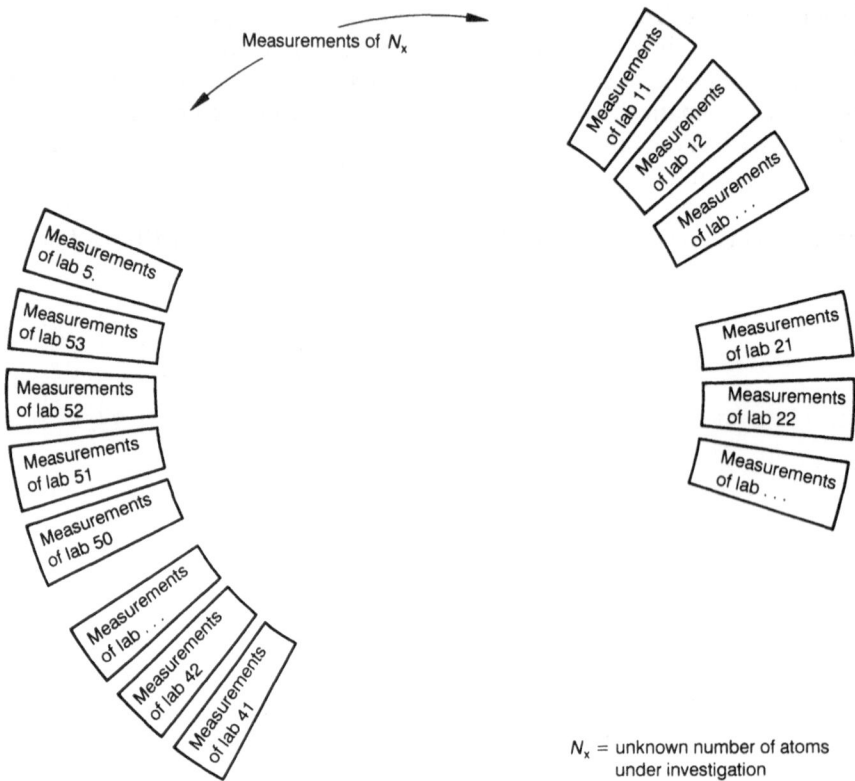

Fig. 4. Isolated measurements: no comparability

comparisons of CRMs will have to become more systematic in terms of amount of the main substance (they have started in the case of gas mixtures which do form relatively simple, almost physical rather than chemical, systems). One must actually conclude from the formal viewpoint, that results of chemical measurements, do lack the required organized basis for being comparable. At present they are declarations of somewhat isolated figures, symbolically pictured in Figs. 3 and 4. Stated another way: they lack comparability because they lack (formal documentation of) traceability. With more and more important decisions being made on the basis of chemical measurements (medical decisions on the basis of clinical measurements, regulatory environmental decisions on the basis of measurements of polluting/toxic substances, food-control on the basis of toxic substance content), the question arises whether this comparability must not be based on a better formal documentation of traceability structure using (legal!) SI base or derived units [3].

To the proponents of traceability of analytical measurements to mass, i.e. to the kg, it should be explained that converting mass to amount is easily done by means of atomic/molecular weights. It is far more logical to do this conversion rather than the reverse (amount to mass). In addition, it implies verifica-

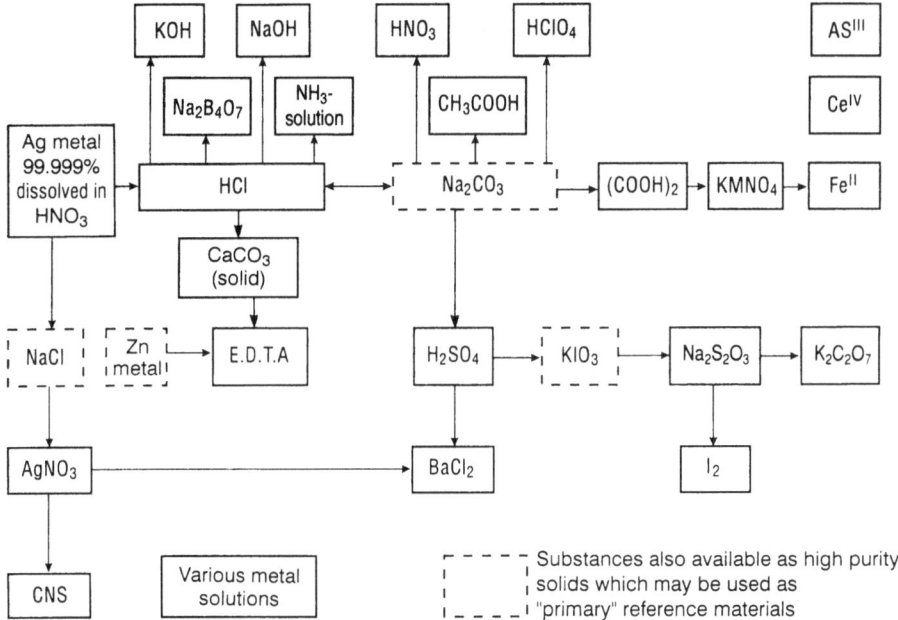

Fig. 5. Scheme relating volumetric solutions to silver chosen as "primary" reference material

tion of the isotopic composition, something which must become compulsory for top-primary CMs anyway now that more natural and man-made variations in isotopic composition are known. The number of them is increasing and the uncertainty on this isotopic composition (or atomic weight) must be included in the overall uncertainty budget. This uncertainty, however, is insignificant relative to normal uncertainties in chemical measurement.

As already mentioned, only chemical measurements which *quantify* can be covered under the SI system e.g. amount measurements or measurements of reaction rates [which are quantitative]. It is difficult to see why elucidations of molecular structure, exploring mechanisms of chemical reactions, etc. would need "traceability" and hence would have to come under the SI system: they are *qualitative* descriptions. The present generalized use of the term "chemical measurements" is too equivocal and needs clarification. Is a qualitative "structural analysis" a chemical measurement? Probably not. But it is if a direct molecular weight of a large molecule or molecule fraction is determined ("measured") and in that case it must be traceable to the atomic mass unit μ used with atomic and molecular masses. The non-quantitative chemical measurements will not be further considered here: it is a topic in itself.

Sometimes schemes for "traceability" in analytical chemical measurements are referred to as shown in Fig. 5 where several "primary" chemical RMs, in the form of (ultra-) pure substances, are interlinked by well known, quantitative high precision – high accuracy chemical reactions as available in titra-

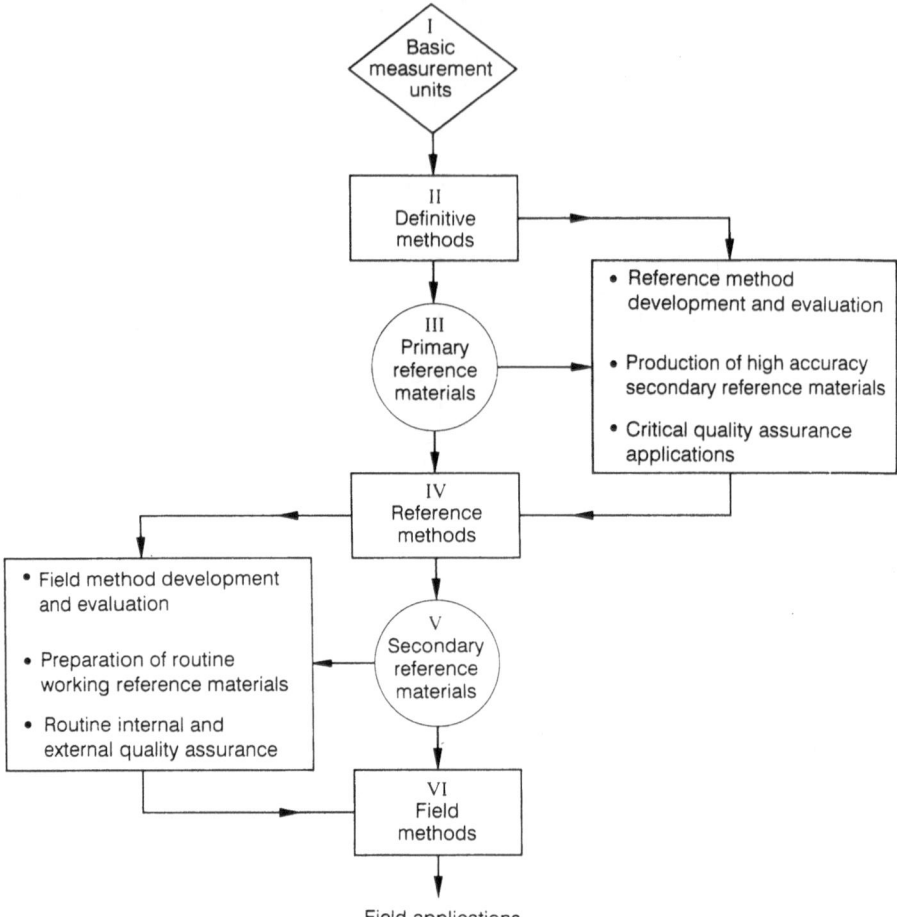

Fig. 6. Scheme illustrating the relationship among the different technical components of an "idealized" accuracy-based chemical measurement system [4]

tions or precipitation processes. It is then said that a number of chemical reactions can link other (less pure) substances to such a system. The "links" which in such a traceability system are constituted by well known chemical reactions, make the quantitative measurement of other substances "traceable" to a limited set of primary RMs. The present situation is that a number of chemical laboratories and producers do provide such pure substances, but there is no organized and internationally accepted system, defining what is (to be) traceable to what. An attempt for another scheme is given in Fig. 6 [4] and yet another one in Fig. 7 [2, 5]. No systems however, have really been implemented although the scheme in Fig. 7 is about to be introduced in ISO guides [6] and a somewhat consistent system was introduced by and in the clinical measurement community (IFCC) in some parts of the world (USA).

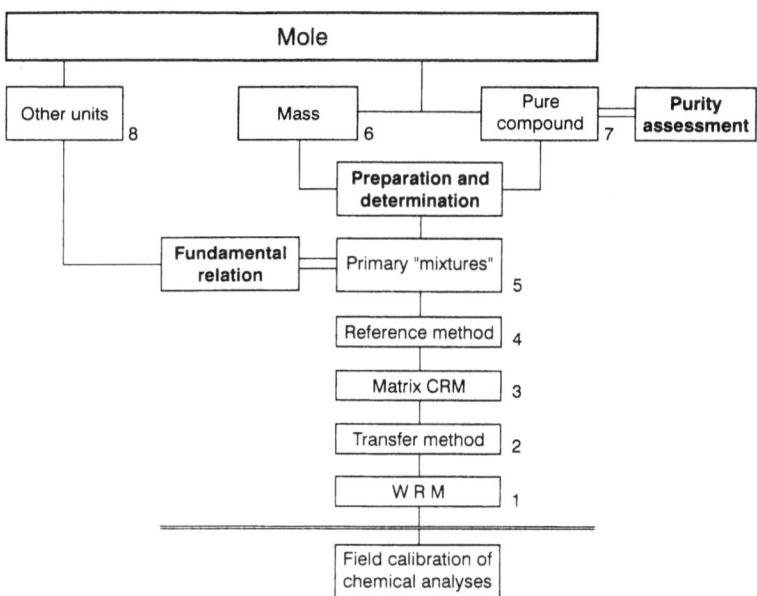

Fig. 7. Traceability in analytical chemistry – flexible general scheme [5]

In Europe, a new association of analytical chemists in EC and EFTA countries (with observers from Eastern Europe, USA) was set up (1989) under the name of EURACHEM specifically to work on the establishment of comparability and of better quality of chemical measurements. This resulted in a worldwide initiative (1993) to pursue the matter: CITAC (Cooperation of International Traceability in Analytical Chemistry). Also the European Association for Metrology, EUROMET, has set up a new subject field "Amount of Substance".

The CIPM has now become active in this matter and, in response to many requests, has agreed to initiate a few selected measurement rounds (trace elements in water 1991–1994, gas mixtures 1993–1995) between a few metrological laboratories as a preparative move to setting up an international system for traceability of chemical measurements. Use is thereby made of a "primary reference method" [1, 7]: isotope dilution mass spectrometry (IDMS) and of high accuracy gas mass spectrometry (HAMAS). In September 1994, the International Committee on Weights and Measures, CIPM, set up a Consultative Comittee on Amount of Substance, similar to the Consultative Committees on the definition of the Metre (1952), of the Second (1956) of Mass (1980), etc. It met for the first time in April 1995. It considers IDMS as particularly suited because of its ratio-of-number-of-particle-counting capability, its species-independence (ions i.e. electrical charges are measured) and its matrix independence: a ratio of numbers of isotopic atoms in the unknown sample is compared to a known added number of isotopic atoms and this ratio is not influenced by the chemical separation process prior to measurement. It is a direct realisation of Eq. (1) in Sect. 7.2 and has therefore great potential [1, 7]. Other methods have also been identified as "primary": coulometry, gravimetry, titrimetry. Similarly, CIPM has set up a Working Group on the Avogadro Constant, advisory to the Consultative Committee on Mass, which started its work in March 1995.

7.6
The "Intersection" Points in a Traceability System

Reference materials are quoted very often as being of the utmost importance in chemical measurements and it must be expected that any international discussion on traceability will largely revolve around "certified" or "primary" or "internationally recognized/accepted" reference materials. It is therefore important to clearly see and agree the correct function of reference materials in any traceability scheme such as shown in Fig. 8. Will *materials* (reference materials) be at the "intersection" (the junctions between radii and concentric circles in

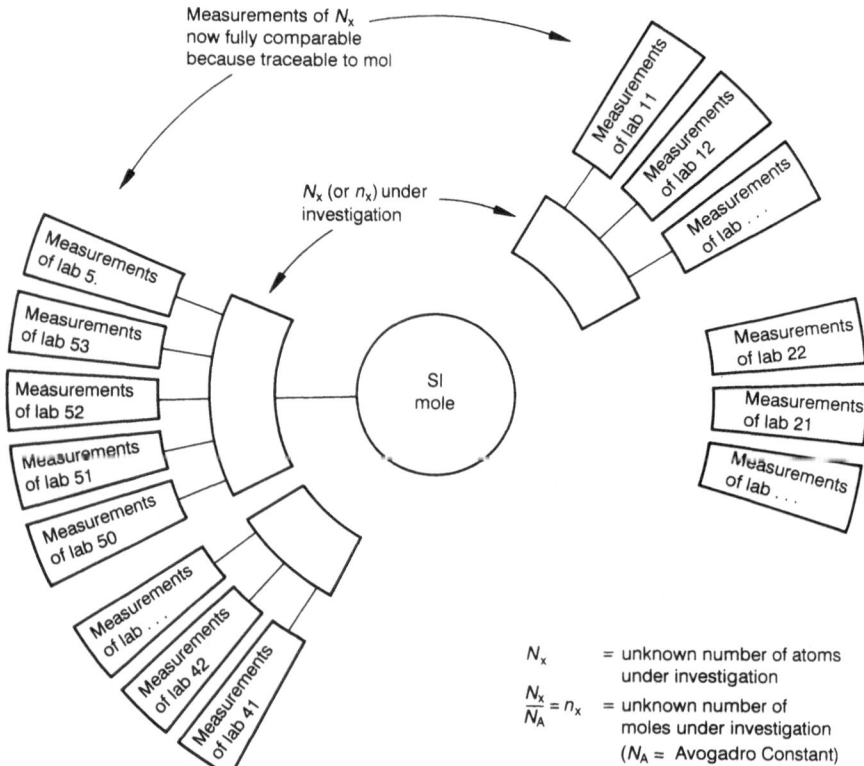

Fig. 8. Traceability to the Mole as support for International Measurement Comparability

Fig. 8) of the traceability scheme, or will *numbers* (reference measurements) form the junctions of the system? It is important to make the distinction.

7.6.1
Are Reference *Materials* at the "Intersection" Points?

Chemists have for decades been looking to RMs as materials for pragmatic, sometimes empirical, corrections of their measurements. In the framework of a discussion on traceability, it is often implied that a RM as "material measure" is a "transfer device" of a number such as in physical measurements where the term "transfer standard" is often used to designate a device (e.g. the national kg) which transfers the knowledge of 1 kg – i.e. a number – from the international kg to, say, a subnational kg. In the organization of international traceability of mass measurements, "weights" are called "transfer standards", they transfer *numbers*. The role of RMs, and especially "primary" RMs, is often similarly interpeted as one of being carriers of numbers *certified* in a "*certificate*". Their aim is then seen as enabling us to make sure that our measurements – comparison of numbers – refer to a known, certified number made available in a RM.

It is important to realize a fundamental distinction between transfer standards/devices in the traceability systems of length and mass which are *extrinsic* properties of a material and amount or amount concentration, which are *intrinsic* properties of a material: intrinsic properties (amount, concentration) of RMs are not analogous to extrinsic properties (mass, length) of national/ subnational metres and kilograms as transfer standards/devices. Thus one cannot use "traceability to a single crystal of $\{N_A\}$ atoms of Si", as a realization of "traceability to the mole", a reasoning almost identical to the one describing traceability of weighings to the national kilograms, then to the international kilogram.

In order to better understand the proper role of RMs, let us examine a little closer how they are – or should be – used in practice.

7.6.2
How Are RMs in Fact Used in Practice?

Since any measurement is subject to a combination of systematic errors, it must be corrected through a "correction" process. There are two ways of carrying out such a process:

1. correct for the errors of every single step in the measurements process as well as for each function of the measurement instrument and include the uncertainty of each correction in the total uncertainty budget of the measurement. This is a very long and cumbersome process which is impractical in daily measurements.
2. a material of almost the same composition, both in terms of the matrix as well as of the concentration of the compounds of interest, and with a known

(certified) concentration, is measured and an overall empirical correction factor k determined for the main suspected error(s) (such as a chemical separation step). Thus a correction factor k is determined by means of a RM:

$$k = \frac{n(RM_{cert})}{n(RM_{obs})} \tag{3}$$

This empirical correction factor is then applied to the measurement $n(X_{obs})$ of the unknown amount $n(X)$:

$$n(X) = k \cdot n(X_{obs}) \tag{4}$$

This procedure may be very practical, but less "metrologically acceptable".

NOTE: It is useful to remind oneself at this point of the original meaning of "calibration", a word which is (mis)used so much when actually "correction" is meant. Calibration originally was used to mean a process or preparing a set of samples (e.g. solutions or set of reference materials or successive dilutions of a reference material) at different concentrations called calibrants, by means of which a "calibration" curve (linear or other) was made which related known values (because "made-up") to the output signals of a measurement procedure for these calibrants. The signal obtained on an unknown sample could thus be converted to a "calibrated" value by an interpolation via the calibration curve. A similar example can be found in the case of radioactivity measurements using calibrants. Unfortunately the VIM definition (6.11) of calibration has become more vague [8]. It is useful to note that calibration applies to a measurement *procedure* whereas a correction applies to a measurement *result*.

A first observation regarding this procedure, is that in many cases no ideal or even good RM is available since those available are not identical enough to the real life sample: the matrix of the RMs cannot be equated anymore to the matrix of the real life sample. It is a fact about which chemists themselves often complain. This situation will get worse: a greater variety of more complex real life samples will continue to appear but the combined production facilities of the world cannot keep up making enough RMs fast enough for all of these cases. The use of "inappropriate" RMs is almost never considered as being the cause of a large contribution to the uncertainty of k thus leading to a too small uncertainty statement on k in Eq. (4). *The "inappropriateness" of RMs must have as consequence a fairly large increase of the uncertainty statement of the measurement in general* and of k in particular. Ignoring this uncertainty contribution may very well be one of the reasons why measurements on the same material by different laboratories are discrepant: their too optimistic uncertainty estimates do not overlap. In fact, the conclusion might well be in a number of cases that the "correction" procedure is invalid.

There is a second observation. When repeated determinations of an empirical correction factor is intended by means of an RM, the spread of these experimental determinations is merely a measure of the spread of various measurement parameters at that moment, not of the value of the measurand. This leads to the conclusion that *the uncertainty of an empirical correction factor* as described above *cannot be interpreted as an uncertainty of the value of the measurand and, therefore, even less as a confidence interval around this value.* If anything, such a confidence interval is almost a statement of lack-of-confidence since it merely describes a spread of the measurement parameters during the process of measuring a correction factor by means of a RM, hence not even at the moment of measuring the unknown sample! This has far-reaching consequences: determining empirical correction factors with their uncertainties by means of inappropriate RMs is not a very metrological procedure to approach the real value of the measurand. It can only be used as a rough approximation of the search for this value. It requires professional assessment by an experienced analyst of a type B uncertainty (not statistics!).

There is a third observation which ought to be made: the fact that the number carried by an RM is known, results in many cases to working towards that value: eliminating selectively measured values ("outliers") which deviate too much, is in fact "optimizing" if not straightforward bringing the correction factor closer to 1 with the then implication of only small "systematic errors" or even of their absence. It is almost invariably forgotten that the correction factor k in Eq. (4) has always an uncertainty since it is experimentally determined through Eq. (3). This is also the case if this factor is equal to exactly meaning no correction and apparent absence of systematic errors. In fact, the uncertainty in the latter case cannot be different from the case where k is different from exactly.

If RMs then are not entirely suited for this overall correction procedure, what is their proper role?

7.6.3
The Real Role of Reference Materials: *Validation*

With their values known, reference materials are eminently suited [17]

a) to validate a measurement procedure: i. e. to give us assurance that the procedure, which should include chemical preparation of the sample, as well as the data acquisition and the data treatment ("software") (see Fig. 9), are suitable and perform properly;

b) more particularly and, if possible, to separately validate a measurement instrument, a basic part of a measurement procedure, (see again Fig. 9) and give us assurance that the instrument is functioning properly, reproducibly, that it is stable, etc. all characteristics that should be compared to previous performances in order to establish this assurance (see part "measurement instrument" in Fig. 9);

c) more particularly and if possible, to separately validate the "data acquisition" and "data treatment" of a measurement procedure (see again Fig. 9);

d) to act as a "validator" of any later improvement of a measurement procedure or of a measurement instrument.

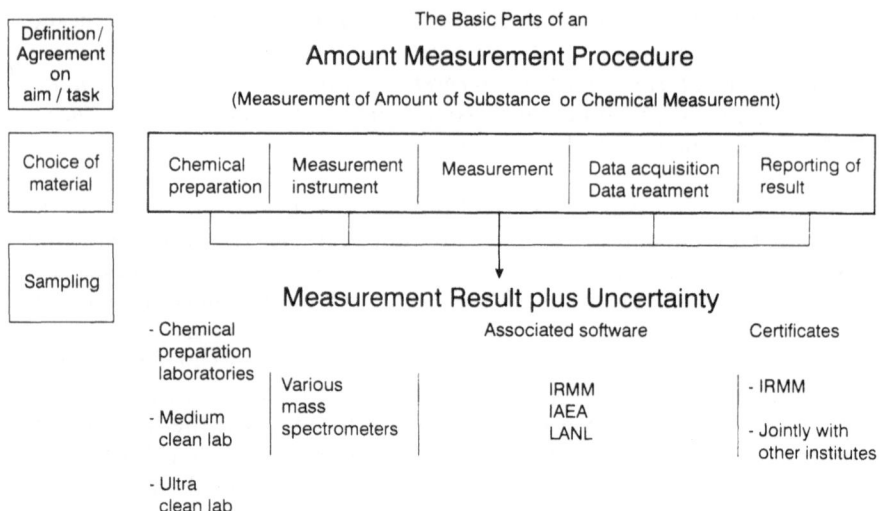

Fig. 9. The Basic Parts of an Amount Measurement Procedure (Measurement of Amount Substance or Chemical Measurement)

This validation process is required *prior* to measuring an unknown sample. It also enables us to (re-)determine inherent uncertainty contributions from the measurement procedure which will always be present in the measurement of real life samples as a part of the latter's total uncertainty. Having this inherent uncertainty at hand, it will now enable us to estimate the additional uncertainty brought along by the measurement of the real-life sample (independent of its homogeneity which is a separate issue).

The important conclusion lies at hand that a validation process (as described above) and a correction process, as described in Eq. (4), by means of a RM, are two different things. This conclusion must increasingly be drawn since in more and more cases, RMs do not have – and in many cases cannot have – the same matrix as the real life sample. This "inappropriateness" of an RM to "correct" (or "calibrate" in the wrong terminology) the measurement of the real life sample, thus leads to the need of distinguishing "validation" from "correction". If a RM is nevertheless used to correct a measurement of an unknown sample – because of the lack of any better means available – a fairly large uncertainty must be estimated by the responsible analyst for the correction factor k, rather than implying that $k = 1$ (ignoring the problem) and has no uncertainty. The question is raised here as to whether the development of "Reference Measurement" procedures, adequately applicable to the real sample matrix is not a more desirable development in the years to come, than trying to make thousands of matrix RMs. Such "Reference Measurements" by specialized chemical reference laboratories on a sample of the analyst's real-life material, could certify the species under investigation in the real-life matrix itself, thus turning the real-life material itself – residing at the place of the analyst – into a highly appropriate RM at the place of the user: an "in-house RM" certified by an external authority. It is recognized however, that such a concept still needs further discussion.

7.6.4
Are Reference Measurements at the Intersection Points?

In a scheme such as shown in Fig. 8, the measurement (i.e. a number) of an unknown sample is traceable to another measurement i.e. another *number* of the same entities (atoms, molecules or species). Numbers resulting from ever more reliable measurements – Reference Measurements – are at the intersections of radii (the "traces" of traceability) and concentric circles in Fig. 8. This is illustrated in Fig. 10. If the reference numbers are certified in SI units (mol) they would automatically be a fraction or multiple of the Avogadro Constant. Ultimately, traceability to the mole (Fig. 10) would then mean traceability to a number (N_A) and an amount measurement in the field would simply be a reliable fraction or multiple of N_A i.e. of a mole of the given entity. Such a definition of traceability would simply mean that we are determining a *reliable number* (of atoms, molecules, ...) in our amount measurement. In terms of measurement technique, this means approaching as closely as possible the "absolute" counting of a specified entity. And that is, after all, the whole purpose of analytical chemistry.

If we now turn back to the traceability discussion, we had concluded that establishing traceability of a measurement of an unknown, was, in fact, the demonstration of a link between an unknown number $N(X)$ of an entity and a known number, N_A (in mol^{-1}). This is explicit in the simple equation

$$\frac{N(X)}{N_A} = n(X) \tag{5}$$

where $n(X)$ (in mol) is the amount of substance. Is measuring an amount of substance in mol, therefore, not automatically ensuring traceability to a fundamental constant, the Avogadro constant, within measurement uncertainty?

7.6.5
The Place of Reference Materials in a Traceability Scheme

From all of the above and from Fig. 10, it results that measurement results (*numbers*) of increasing quality (smaller uncertainty) link the measurement of a real life sample to a level of higher metrological authority (only as high as needed), which means to one of the more inner circles of Fig. 10. The measurement processes needed to do that – link one circle to another – must be validated by RMs. Their role seems to be to validate the "bridges" between the circles in Fig. 10. An attempt to illustrate this, is given in Fig. 11 which, if superimposed on Fig. 10, shows the RMs in the "bridging" roles between the concentric circles. From this picture it is also clear that the nearer to the centre the RMs perform their function, the better their quality must be (more "primary"). Of course, RMs also must be traceable to SI for the entire system

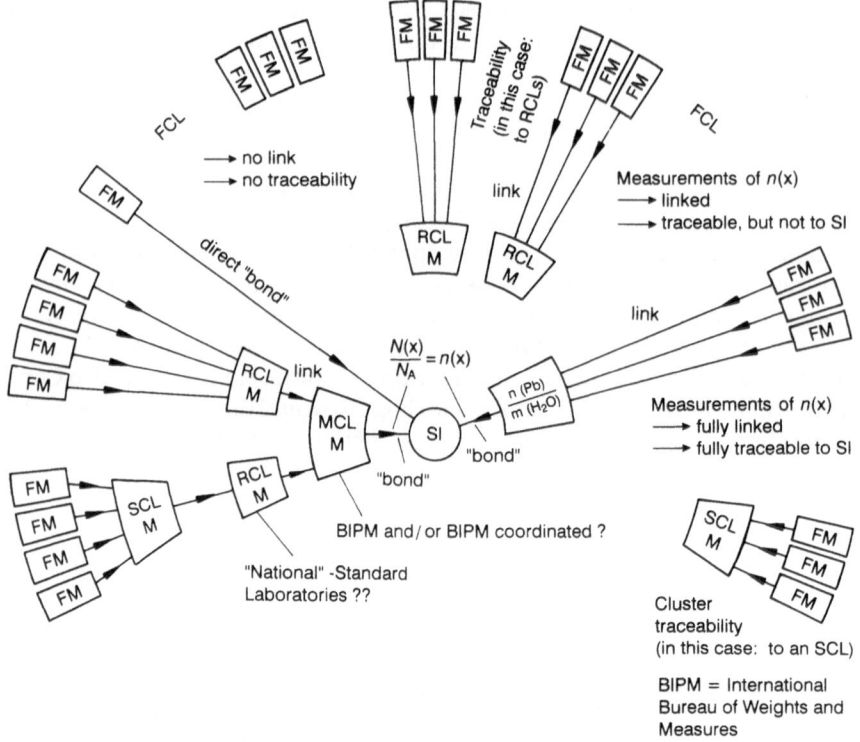

Notes: 1. Traceability implies a relationship usually with a direction (arrows)
         towards higher authority (in metrology, not in specialized chemical know-how)
      2. The inverse of relative uncertainty is a measure of
         reliability or link strength
      3. An RM is a validator of an instrument and/or a method used
         (for the intrinsic property: amount of substance) prior to a measurement

Legend: 1. $N(x)$ = unknown number of entities to be determined

$\dfrac{N(x)}{N_A} = n(x)$ = unknown amount of substance under investigation

2.
Field	Sectorial	Reference	Metrological
Chemical	Chemical	Chemical	Chemical
Laboratories	Laboratories	Laboratories	Laboratories
= FCL	= SCL	= RCL	= MCL

Fig. 10. Attempt to structure Traceability to SI of Measurements of Amount of Substance: Field Measurements (FM) to SCL Measurements, RCL Measurements, MCL Measurements

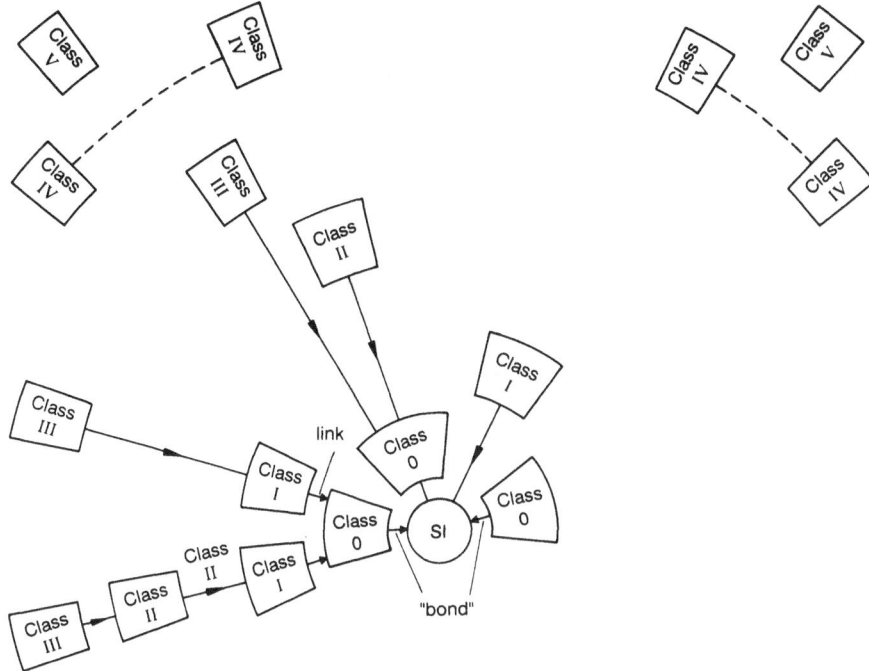

Table: Proposed classes for reference materials
in terms of degree of traceability to SI

Class	Description and criteria in terms of traceability to SI
0	Pure specified entity certified to SI at the smallest achievable uncertainty
I	Certified by measurement against class 0 RM or SI with defined uncertainty by methods without measurable matrix dependence
II	Verified by measurement against class I or 0 RM with defined uncertainty
III	Described linkage to class II, I or 0 RM
IV	Described linkage other than to SI
V	No described linkage

Fig. 11. Attempts to structure Traceability of Reference Materials, certified for Amount of Substance, to SI [19]

to be consistent. This is shown in Fig. 11. By no means, is it implied here that the uncertainties discussed are the main contributors to the total uncertainty of a measurement. In many cases much larger uncertainties occur at an early stage of the measurement process e.g. in the preparation of the sample for measurement. They also have to be studied and added.

7.7
Purposes of Traceability of Amount Measurements

Chemical measurements are flooding our society and literally millions of measurements are done each day in a community such as Europe or the USA. Some of them are important because important decisions are based on them. It is a measurement around a "decision level" which determines whether water is legally drinkable or not (Fig. 12). It is a measurement around a "decision level" which determines whether food must, for example, be taken away from commerce because of the Cd content. Regardless on whether the decision level is aptly and appropriately chosen and legally or regulatory fixed at a given level (whatever that level), a measurement must determine whether a material characteristic complies with that level. The measurement has important consequences and must hence be demonstrated to be reliable and traceable to the scientifically accepted – and legal – international SI system as much as possible. Measurements between two or more parties (shipper-receiver, inspector-inspected, country to country) must have a solid basis for their comparability[3].

Having been "decreed to be comparable" (as in the case of "Europe 1993") is just not sufficient. One of the most important bases of comparability is demonstration of traceability to a recognized and legally agreed international measurement system – as explained above. *"The primary and most general purpose of traceability is to ensure that measurements are what they purport to be"* (G. Price, personal comment).

It has been stated in the literature that "considerable evidence exists in the literature that few analytical chemists pay attention to the question of the reliability of the analytical results they produce. These chemists believe that a natural law exists in measurement science that if the directions for conducting a measurement are followed, the true value necessarily results" [9].

Is indeed more attention paid to precision or reproducibility rather than to accuracy? Could there be a naive and unscientific belief that the results of fully automated (and therefore non-transparent) measurement instrumentation with associated "black box" software, are automatically correct? Has this virtually eliminated the personalized responsibility of the analyst for the reliability of the *number* he/she is supposed to deliver? Are chemists indeed underestimating the uncertainties of their measurement processes because of the complexity of their science?

---

[3] some use terms as "compatibility", "consistency", "harmony"; we use "comparability" when measurements of the same characteristic or property are quantitatively "compared" with each other.

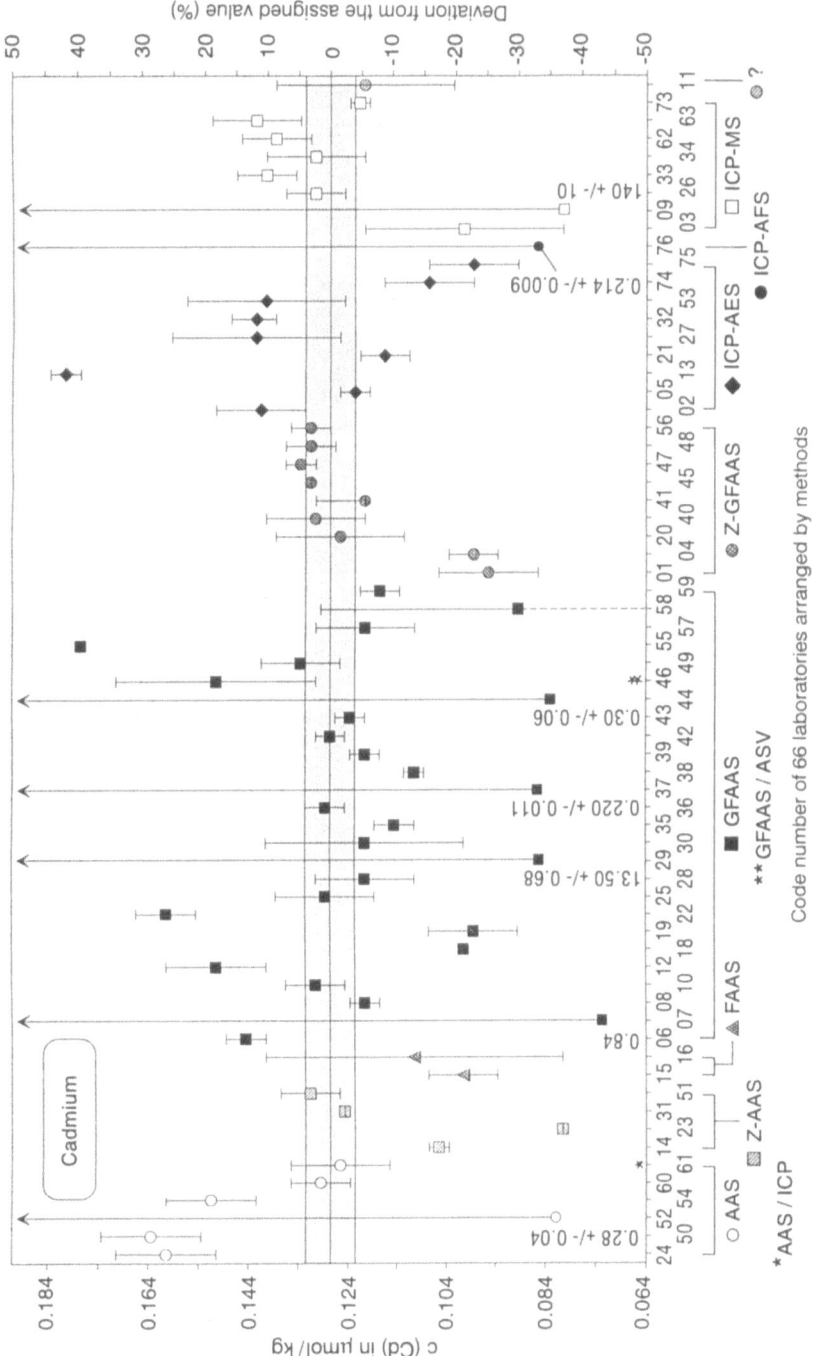

Fig. 12. Results of Cd measurements in IRMM's International Measurement Evaluation Programme IMEP, Round 3 Trace Elements in Water

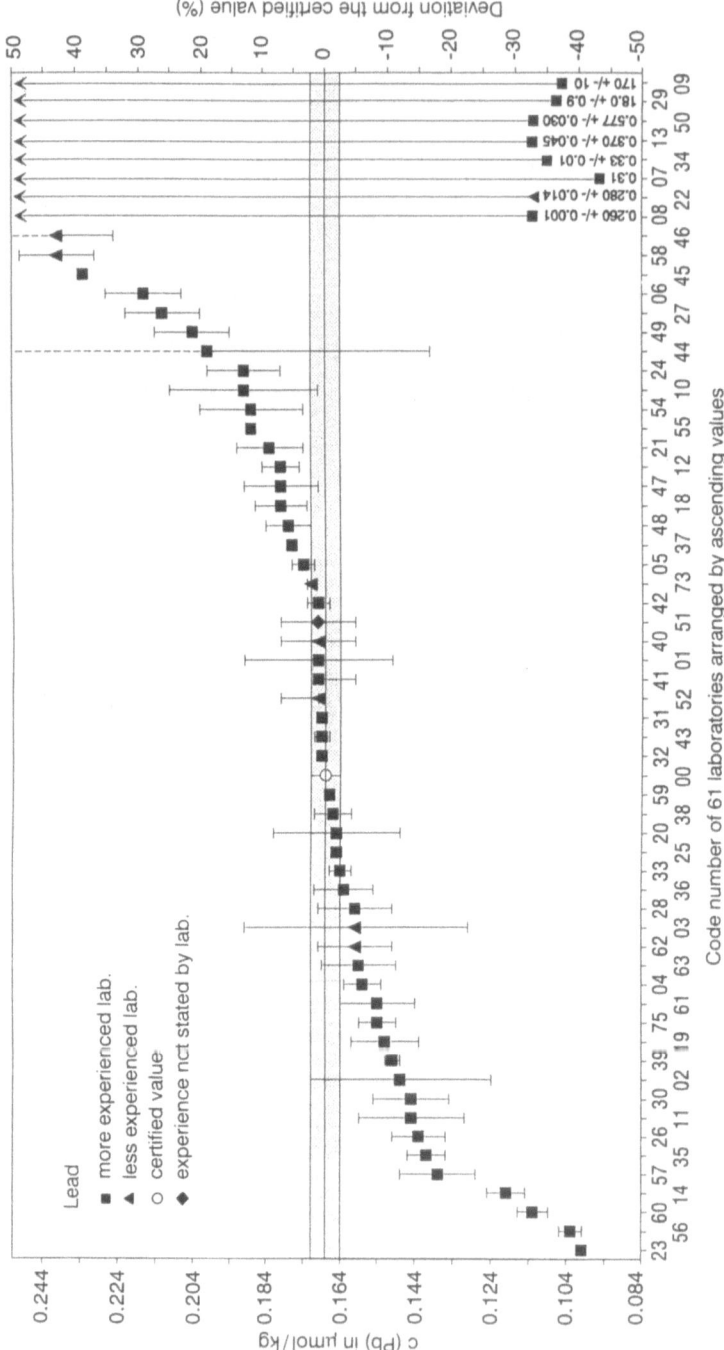

Fig. 13. Results of Pb measurements in IRMM's International Measurement Evaluation Programme IMEP, Round 3 Trace Elements in Water

It must be the purpose of any traceability scheme to provide comparability, compatibility and consistency between the huge numbers of chemical measurements needed everyday everywhere. This is required in order to give the parties concerned, the authorities involved and the public at large, the assurance that rather than being statements of isolated figures, measurement results are related to the entity that they claim to measure, that they do this in the same unit everywhere so that they are directly comparable within the limits of their quoted uncertainty, even for an unqualified person, and that they are naturally and logically based on an approriate scientific/technical structure of international nature.

Let us now attempt to fix criteria for the specific case of traceability to the mole.

7.8
Criteria for Traceability of Amount Measurements to the Mole:

Some definitions for traceability have been published in recent years.

1. **The VIM definition (6.10) of Traceability reads [8]:**

   "property of the result of a measurement or the value of a standard whereby it can be related to stated references, usually national or international standards, through an unbroken chain of comparisons all having stated uncertainties".

   Notes: – the definition does not mention SI: traceability is to a system (e.g. SI) or to agreed artefacts (kg, reference materials);
   – the definition is usable in cases where there are no SI units, or not even derived unites (e.g. hardness, flash point, octane number),
   – the definition leaves "an unbroken chain of comparisons" undefined; ongoing intercomparison programmes probably play a very important role in the "chains".

We cannot escape the feeling that the time has come to take the VIM definition of traceability quoted in Sect. 7.8.1 a step further, taking into account the views generated by the concept of "amount measurement" and by the application of the current definition to the realization of the volt and the ohm through the quantum Hall and the Josephson effects.

2. **The new definition by ISO of a Certified Reference Material requires "traceability to an accurate realization of the unit in which the property values are expressed" [10]**

   Notes: – the inclusion of a "reliable realization" is a considerable progress (see description of some other – physical –quantities in Sect. 7.4);
   also the inclusion of "the unit in which the property values are expressed" is an improvement and very relevant as may have become

clearer from Sects. 7.2 and 7.3 for the unit mol and the unit kg and by extrapolation for derived and – even empirical – units in use;

- if the concept is correct that traceability of measurement *results* is what we are really trying to set up, then traceability of measurement results from one laboratory to a laboratory with higher measurement authority, should be included in the definition; then it follows that QA and QC programmes, external as well as internal, become part of the definition; this would contribute to a better understanding of the operational meaning of traceability, which is badly needed.

3. Four possible definitions are mentioned by B.C. Belanger [11]

a) Traceability is the ability to demonstrate conclusively that a particular instrument or artifact standard either has been calibrated by NBS at accepted intervals or has been calibrated against another standard in a chain or echelon of calibrations ultimately leading to a calibration performed by NBS.

b) Traceability to designated standards (national, international, or well-characterized reference standards based upon fundamental constants of nature) is an attribute of some measurements. Measurements have traceability to the designated standards if, and only if, scientifically rigorous evidence is produced on a continuous basis to show that the measurement process is producing measurement results (data) for which the total measurement uncertainty relative to national or other designated standards is quantified.

c) Traceability means the ability to relate individual measurement results to national standards or nationally accepted measurement systems through an unbroken chain of comparisons.

d) Traceability implies a capability to express quantitatively the results of a measurement in terms of units that are realized on the basis of accepted reference standards, usually national standards.

Notes: – the first definition is not quite acceptable as such on the international scene, but may provide a useful USA domestic definition; it carries the notion of traceability to a laboratory which may or may not be useful;
- the second definition focusses much more on the quality of the result and its uncertainty thought of as a sort of "total uncertainty" including the uncertainties of each step along the traceability chain to designated "standards"; in doing so it stresses the importance of the quality of the result and goes along the lines set out in Sect. 5: traceability of numbers;
- the third definition also points more to results (numbers) rather than to reference materials at the key points of any measurement system; it does not mention "calibration" either;
- as for the fourth definition, it is rather vague and leaves questions open on the "realization of units on the basis of accepted reference standards"; also "usually national standards" is not acceptable for international use.

4. In our opinion, the best definition so far of traceability has been given by Nicholas and White [12]:

"Traceability is the ability to demonstrate the accuracy of a measurement in terms of the SI unit". We would like to amend it and propose the following: "Traceability is the ability to demonstrate the result of a measurement and its uncertainty in terms of the SI unit". This definition carries the notion of

1. ability
2. demonstration (to oneself and to others)
3. uncertainty
4. "accuracy" as used in the past including the connotation that the whole aim of the SI system is simply to carry out accurate measurements. When is a measurement result accurate? Answer: when both the result and its uncertainty are described in SI units.

The basic concept of *amount of substance* stems from the particulate nature of matter. Due to chemists' requests, this led to amount of substance becoming one of the seven base quantities of our International System (SI) of measurements (in 1971). Thus, it follows that one of the seven base units of the International System of Units is for amount of substance, namely the mole. It was defined as follows (14th CGPM 1971):

"The mole is the amount of substance of a system which contains as many elementary entities as there are atoms in 0.012 kilogram of carbon-12. When the mole is used, the elementary entities must be specified and may be atoms, molecules, ions electrons, other particles, or specified groups of such particles".

An attempt is now being made to formulate criteria for "Traceability of (Chemical) Amount Measurements of the Mole" on the basis of the following explicit expression of the VIM definition of Traceability, for application in metrological chemical matters: "Traceability of a measurement is the ability to trace the result of that measurement back, with an acceptable uncertainty, to the closest form of an accurate realization of the relevant SI unit (basic or derived) for the quantity measured, or – failing these – of a relevant internationally recognized empirical unit for that quantity".

The criteria of traceability to the mole are logically derived from the above:

1. The traceability concerns a specific, defined entity.
2. The traceability applies to a quantitative amount-of-substance measurement of verifiable total uncertainty.
3. The measurement of the amount of substance is conducted by the best possible method for counting entities.
4. The result and its uncertainty are expressed in mole units.
5. Realization of traceability is by *described* measurement procedures and instruments.

Explanatory Comments

1. Traceability to the mole for amount measurements is held to be traceability to SI. It is equivalent to traceability to N_A, the Avogadro constant.

2. When amount measurements are performed by methods which are not based on the particulate nature of matter (e.g. gravimetric determination by the weighing of a stoichiometric compound), traceability to the mole can be established by the addition of appropriate steps (conversions to amount) with their corresponding uncertainty; it is obvious however, that this lengthens the traceability chain by adding links which must be evaluated for the additional uncertainty each link contributes.
3. The relation of numbers of entities in chemical reactions and to energy conversions involving substances is direct and favours measurements by the mole over measurements by the kilogram.
4. Commonly chemical measurements involve that of a specific entity in a solution or mixture of possibly unknown chemical entities; it is more transparent and logical to use the mole unit for the measurement of entities; it is convenient to measure the mixture of chemical entities by mass or volume.
   One thus could use mole (of defined entity) per kilogram (of material): $mol \cdot kg^{-1}$, in measurement results, not %, ‰, $kg \cdot kg^{-1}$, ppm, ppb, etc.
5. We have here introduced a more specific meaning of 'substance' as opposed to the more general and intuitive popular understanding of the term.
   If, say, we have an amount of $n_1(O_2)$ and add an amount $n_2(N_2)$, we do not have an amount of $(n_1 + n_2)$, as might popularly be imagined (since the substance is not specified).
   [This is not very different from the old truth that you cannot compare (or add) apples to oranges, but p_1 apples plus p_2 oranges do make $(p_1 + p_2)$ pieces of fruit. Or, pentoses and hexoses comprise the naturally occuring monosaccharides].
6. If traceability of an amount measurement is claimed to other than the mole unit itself through an instrument, material, procedure or standard, then that other intrument, material, procedure or standard must be credibly described and their relation to the mole clearly established.]

7.9
How Can Traceability to the Mole Be Established?

Attempts to offer a realization of a measurement traceable to N_A are tested out in IRMM's "Regular European Interlaboratory Measurement Evaluation Programme" REIMEP for nuclear materials (U, Pu) and its "International Measurement Evaluation Programme" IMEP (for other materials) (Figs. 12–13) [13, 14]. Unknown samples are sent to interested laboratories. These laboratories return one value as a result of their measurements and are asked to make a statement of an "uncertainty claiming to contain the true value" (a very useful operational definition of accuracy). The closest-to-the-truth value is established by isotope dilution mass spectrometric measurements (IDMS) [1, 7] which are traceable to the mole: the amount measurement procedure used to fix them is a "primary method of measurement" as now defined by CCQM [15].

Its basic underlying philosophy is that the best value to serve as a reference, is obtained from measurement processes and their uncertainties which are fully understood, not from averaging a number of values from laboratories. An important precaution in REIMEP and IMEP, is that the value traceable to the mole, is only released after all participants have submitted their measurement results. A known sample such as an RM can only serve as a validator of method and instrument and is not very helpful for making a trustworthy picture of the real State-of-the-Practice of measurements of unknown samples: "analysts exhibit better precision and less bias when analytical samples are analysed as known rather than as unknowns" [16–18]. This remark discourages putting natural matrix reference materials at certain key points in any traceability system because they have known (certified) values whereas we need to demonstrate that *the results of the measurements of unknowns are traceable*. The idea emerges again that what we really need is *traceability of measurement results in the field (which are numbers) to a number (N_A)* (Fig. 10): a picture for traceability of chemical measurements emerges which is the same as arrived at in Sect. 7.6 where *bridges* related a number in the sample measured to another number and ultimately to N_A by means of a calibrated measurement (these bridges are the *lines* along the radii and between the junctions in Fig. 10). In this picture the purpose of RMs with known values is to correct the "bridging" measurements.

Such a model also enables us to distinguish between traceability and uncertainty. Traceability is the ability to trace a result back to something. Traceability has to do with the *links* (traces) in the system: the radii in Fig. 10. Uncertainty has to do with the strength of these links: very tight e. g. in precisely corrected measurements with small measurement uncertainties; or rather loose: e. g. when uncertainties are large and the measurements possibly even uncorrected. However, even in the case of larger uncertainties, the link (a "*trace*") must be there when demonstration of *trace*ability is needed or desired: the ability to find the trace and to conclude on the uncertainty of the trace. A comparison with mass measurements helps: some weighings are not carefully carried out, i. e. with large uncertainties – link or trace is weak – but that is no reason why weighings and weights must not be *trace*able to the national and international kg.

7.10
Conclusions

The ultimate goal of any measurement, any contributing step towards measurement comparability and any concept of measurement traceability clearly must be: to produce results which are accurate and reliable within the stated uncertainty, all uncertainties of all steps in the measurement and the traceability chain having been included. Performing metrology or metrological measurements – including amount of substance – actually means to be constantly *conscious* of the problems of the measurement and of the measurement process. This is a basic duty of anybody measuring.

It seems clear that the idea of traceability is not sufficiently widespread in the field of chemical measurements. To support the current need for global measurement comparability, a clear and generally accepted concept of traceability and an infrastructure to support such a concept, are still missing.

It seems equally clear that the chemical community must soon produce a clear picture for the outside world by means of which it can demonstrate that a logical structure is put in place for measurements in chemistry which are relevant to societal decision-making and that this structure is agreed upon by all and can be demonstrated to function.

A set of rules of "Good Practice" as well as formal procedures for accreditation may be useful and in fact needed. However, mostly such rules only monitor conformity with written procedures. They are "conformity assessments". It may be correct that in a number of applications, two parties feel happy because their measurements on the same material happen to yield nearly the same results by sheer coincidence, or by being both equally wrong. In important cases or on the international scene, however this is not satisfactory in the end, nor would it be accepted in practice. An amount of substance corresponds to an existing physical or chemical reality. It is of prime importance that we know and measure the *real* world. It is only by pursuing this reality and arriving at a reliable statement of a range containing this reality, that we can create a truly satisfactory scientific and societal basis for agreeing on measurement results internationally. It is also a moral obligation for analysts to do so.

To the public at large the assurance must be given that measurements *are* "good" and that they not merely *appear to be so*. This can best be done by demonstration.

Acknowledgements. The author wishes to acknowledge many fruitful discussions with H. S. Peiser, Visiting Scientist at IRMM and formerly associated with NIST, and G. Price of the National Standards Commission (Australia) who provided extremely refreshing considerations on various drafts of this study. Exchange of in-depth comments and views with S. Rasberry and W. P. Reed (NIST) and W. Wegscheider (Universität Leoben) caused considerable reshaping of sections with respect to both substance and wording, and leading to considerable progress in mutual understanding. He is also highly appreciative for frequent exchanges of views with A. Lamberty, K. Mayer, P. Taylor and J. Pauwels, colleagues at IRMM as well as with B. King at LGC. He wants to pay tribute to the work of Mrs. H. Kerslake, who has typed many versions of this study during the time it grew from a first, to a third edition, time and again incorporating improvements as a result of comments from others. Similarly, H. Koekenberg of the IRMM Drawing Office, patiently prepared the many versions of various figures before they actually presented what was described in the text in comprehensive pictures.

7.11
References

1. De Bièvre P (1993) Isotope Dilution Mass Spectrometry as a Primary Method of Analysis, Anal Proc Royal Soc Chem 30 (8):328–333
2. Marschal A (1980) La Mole: de la définition à l'utilisateur, Bull BNM 39:33–39
3. International Union of Pure and Applied Chemistry IUPAC, Quantities, Units and Symbols in Physical Chemistry, Blackwell Scientific Publications, Oxford 1993
4. Uriano GA, Gravatt CC (1977) The Role of Reference Materials and Reference Methods in Chemical Analysis, CRC Critical Reviews in Analytical Chemistry 361–412
5. Marschal A (1995) 2nd Symp EUROLAB, Firenze (April 1994): also Bull BNM in press
6. Calibration of Chemical Analyses and Use of Certified Reference Materials, ISO/REMCO N 262 E/F Nov 1994, prepared by Marschal A
7. De Bièvre P (1990) Isotope Dilution Mass Spectrometry: What can it contribute to Accuracy in Trace Analysis? Fresenius J Anal Chem 337:766–771
8. International Vocabulary of Basic and General Terms in Metrology, ISO, Genève 1993
9. Horwitz W (1992) J AOAC International 368
10. ISO Guide 30:1992 (E/F)
11. Belanger BC (1980) ASTM Standardization News January 22–27
12. Nicholas JV, White DR (1994) Traceable Temperatures p 24. Ed Wiley J & Sons, Chichester
13. De Bièvre P, Wolters WH (Mai 1987) The Regular European Interlaboratory Measurement Evaluation Programme (for U/Pu), Proc 9th Ann ESARDA Symp Safeguards and Nucl Mat Management, London 375–379
14. Lamberty A, Lapitajs G, Van Nevel L, Götz A, Moody JR, Erdmann DE, De Bièvre P (Sept 1993) The IRMM International Measurement Evaluation Programme (IMEP), IMEP-3: Trace Elements in Synthetic and Natural Water, Intern Symp on Nucl Anal Meth in Life Sciences, Praha 13–17
15. CCQM, first Meeting in BIPM, Sèvres, April 1995
16. Thompson M (1989) Proc 3rd Intern Symp Harmonization of Quality Assurance Systems in Chemical Analysis, Washington, ISO/REMCO (184):183–189
17. Banes D (1969) J Assoc Off Anal Chem 52:203–206
18. Horwitz W op cit 370

Further reading may be found in:

19. De Bièvre P, Kaarls S, Rasberry S, Reed WP Metrology and the Role of Reference Materials in Validation and Calibration of Traceability of Chemical Measurements, Proc National Conference of Standard Laboratories (NCSL), Dallas TX (USA) July 1995 and Symp on Total Quality Management, Melbourne (Australia), December 1995
20. De Bièvre P, Kaarls R, Rasberry S, Reed WP (1996) Measurement Principles for Traceability in Chemical Analysis, Accreditation and Quality Assurance ACQUAL, 1 accepted for publication

Chapter 8

Reference Materials for Quality Assurance

Ph. Quevauviller and B. Griepink

8.1
Introduction

Accurate measurements are essential to the functioning of modern society. Without them, industries, particularly high technology ones, cannot operate, trade is impaired by disputes, health care becomes empirical and legislation, ranging from environmental and worker protection to e.g. the operation linked to the Common Agricultural Policy or the Single Market, cannot be successfully implemented. The harmonization of measurement systems is a well recognized need; it may be achieved either by means of directives or norms which, however, does not solve all the problems. Indeed, the analyses required for the implementation of these norms and directives are sometimes so difficult that, even when applying the same method, laboratories may still find different results. It is clear that such a disagreement between laboratories does not allow the norms and directives to be respected and, therefore, these have no harmonization effects. As a consequence, measures to ensure a good quality assurance of analysis were established, involving rules and guidelines (e.g. Good Laboratory Practice, ISO 25 and EN 45000 series etc.) and accreditation systems. Their main goal is obviously to ensure the accuracy of data produced, and hence their comparability. Various means to achieve accuracy exist and are applied in good laboratories [1, 2]; they include the existence of a quality accurance manual, training of personnel, a good managerial structure of the laboratory, the use of validated methods, the existence of a quality assurance manual, the application of statistical control principles (e.g. control charts), external quality control measures (e.g. participation in intercomparisons leading to confrontation with results of other laboratories) and use of certified reference materials. In the limited scope of this chapter attention will be paid mainly to the role of reference materials (RMs) and certified reference materials (CRMs) in chemical analysis. They are necessary for various of the above mentioned prerequisites: method validation (CRM), monitoring of the state of statistical control (RM), samples in intercomparisons (RMs) etc. and thus represent one of the most necessary tool for a good analytical quality control [3]. Other quality assurance aspects (e.g. strategy, management etc.) are developed in chapter 4 of this volume.

8.2
Definitions

The following ISO-definitions are relevant (ISO, [4]):

- Reference Material (RM): A material or substance one or more properties of which are sufficiently well established to be used for the calibration of an apparatus, the assessment of a measurement method, or for assigning values to materials
- Certified Reference Material (CRM): A reference material one or more of whose property values are certified by a technically valid procedure, accompanied by or traceable to a certificate or other documentation which is issued by a certifying body.

8.3
Requirements for the Preparation of RMs and CRMs

The fundamental difference between RMs and CRMs is that some parameters in CRMs are known with great accuracy. In order to arrive at sound conclusions on the laboratory's accuracy, the CRM used for verification should have a similar composition to the unknown sample; a basic prerequisite is that the same sources of error can be encountered in analysing the CRM or the unknown. CRMs of good suppliers are traceable to the basic units and therefore link the user's results to those of the international scientific community.

8.3.1
Selection

The above requirement of analytical representativity means, in most cases, similarity of:

- the matrix composition;
- the contents of the analytes;
- the way of binding of these analytes;
- the fingerprint pattern of possible interferences;
- the physical status of the material.

In preparing a RM these items should be taken into full consideration. For example, for the determination of compounds bound to matrix constituents (e.g. organometallic and organic compounds), artificially spiked materials are not representative for real samples. For practical reasons, however, the similarity cannot always be entirely respected. The material has to be homogeneous and stable in order to assure that samples delivered to the laboratories are the same. Therefore, compromises have to be made and the preparation of the material has to be adapted.

8.3.2
Preparation

The amount of RMs or CRMs to be prepared is a function of the analytical sample size, stability, shelf size and frequency of use; it may vary from 5 to 20 kg of solid material (e.g. soils, sediments etc.) or 5–20 l (water) for the preparation of RMs to be used in routine analysis to up to 100 kg of solid material or several cubic meters of liquids in cases when CRMs are to be prepared. Indeed, the RMs are often used as laboratory materials in one single laboratory whereas the CRMs are produced on a larger scale. In some cases, however, RMs are also produced on a large scale (e.g. for interlaboratory studies). The sample intake necessary and the number of laboratories involved in the analysis of a given material actually dictate the production required. The producer needs to be equipped to treat large amounts of material without substantially changing the analytical representativity. The treatment of e.g. 3 to 5 kg raw material is already the limit for normal laboratory equipment and manual processing; for larger batches and especially larger volumes of material it is necessary to scale up to half industrial size.

Typical operations are: grinding, sieving, filtering, mixing or homogenisation of the materials which can be performed only in specialized laboratories or industries. The most sensitive and difficult step of the preparation is the stabilization; it has to be adapted to each particular case and should be studied in detail before processing in order to respect as much as possible the integrity of the material. Usually the materials are dried to avoid chemical or microbiological changes. This may be achieved by oven-drying (e.g. for sediment, [5] or by freeze-drying (e.g. for fish, depending on the volatility of the analytes and matrix components.

Some materials can be sterilized by γ-irradiation (^{60}Co source). In reality this treatment is mainly possible for materials to be used for element determinations as many compounds decay upon γ-irradiation (e.g. tin compounds, [8], or pesticides, [9]. Freezing of the material is also possible but the resulting material can only be used once as refreezing may not lead to an homogeneous material. Moreover, the worldwide shipment of frozen materials is cumbersome and not always achievable. The material must be homogenized and stored in adequate vials. For gases and liquids, homogeneity is not the most difficult problem; the stability however causes great concern. Solid materials are difficult to homogenize; it has been shown, however, that if (i) the particle size is less than 125 μm, (ii) the particle size distribution of the material is sufficiently narrow and (iii) the density of the particles does not differ by more than a factor of 3–4, the homogeneity is sufficient for sample intakes of less than 100 mg [2]. Therefore a proper grinding procedure and a thorough homogenization prior and during the filling procedure is recommended. Unfortunately, the low particle size presents some drawbacks as it leads to materials which are usually more easy to analyze than real samples (better extractability of analytes or easier matrix digestion because of the large contact surface). In addition static electricity may cause subsampling difficulties for some materials with very low particle size and low water content (as observed e.g. in the case of human hair, [10]).

8.3.3
Homogeneity

A sub-sample of an RM can be used in chemistry only once as it is usually destroyed during the analysis. Therefore, the quantity of material in the bottle or ampoule should be sufficient to perform one or sometimes several determinations; the more sub-samples taken the less the chance that the bulk is still the same.

A verification of the homogeneity has to be performed to ensure that within a bottle or ampoule and from one vial to another the contents are the same (within- and between vial homogeneity). A current practice developed within the Community Bureau of Reference (BCR) implies that the homogeneity of the materials is verified by performing e.g. 10 replicate determinations of the analyte of concern within one bottle (within-bottle homogeneity) and by one determination in bottles set aside at regular intervals during the bottling procedure (2 to 5% of the bottles can be used for this between-bottle homogeneity); the coefficient of variation (CV) obtained may be compared to the CV of the final step of the method (assessed by e.g. 5 replicate determinations of one digest solution). In such a comparison with an extract or digest, that the extraction or digestion procedure attributes to the random uncertainty of a result is not considered; it, therefore, leads only to conservative conclusions.

The minimum sample size for which the homogeneity is sufficient should be verified and given by the manufacturer. Below this intake level, the uncertainty caused by inhomogeneity attributes significantly to the uncertainty of the reference (or certified) values (i.e. maximum 30% of the total uncertainty). An additional problem is caused by segregation during transport and long term storage; special care should be taken for the rehomogenization of the material before taking a test portion.

8.3.4
Stability

The properties of the material and the parameters investigated should remain unchanged over long periods and the long term stability should therefore be studied. The stability can be estimated by evaluating the behaviour of the material under accelerated aging conditions e.g. elevated temperatures. The BCR is currently performing studies of the stability of CRMs at room temperature and at $35-40\,°C$ and to detect a possible evolution by comparing the results obtained with samples stored at $-20\,°C$ (at which temperature changes are very small to negligible). The results obtained at $+20\,°C$ may lead to an assessment of the sample stability at ambient laboratory temperature whereas the results obtained at $+40\,°C$ are used to assess the worst case conditions (e.g. during transport) and allow us to evaluate the stability of the material over longer periods of time; it is indeed assumed that a sample stable at + 40 °C during one year may be stable at + 20 °C for a longer period (Arrhenius). This assumption does not hold in cases of spoilage by certain bacteria or moulds having optimum temperatures for their metabolism at 20 – 35 °C; however, the fact that usually water is removed to a content below 5 % by mass, or that samples are γ-irradiated to reduce the number of bacteria gives confidence that such spoilage does not occur. Stability checks should therefore be carried out at regular intervals (e. g. three years) after certification.

8.3.5
How to Obtain Reference Values

Several technically valid approaches may be used to obtain reference values in a RM [1], one approach consists of using a method in which one single laboratory is proficient and which has given proof of good performance and to determine the content of the elements/compounds of concern along with the analytical uncertainty. It is stressed here that this value is meant to be used for the verification of the long term reproducibility of a laboratory; indeed, if the laboratory has a systematic deviation in using their single method, the value obtained is not accurate. In addition, the use of a single method in one laboratory does not give an accurate estimate of the uncertainty as can be achieved by other methods. However, if a single laboratory has given proof of being able to perform an analysis with great accuracy and precision, the value should be rated highly.

Another approach more often adopted consists of using several independent methods within a single laboratory. The results become more reliable when two or more independent methods give the same results. Consequently the reliability increases further if different methods applied in different laboratories by different technicians give results that agree closely.

8.3.6
How to Obtain Certified Values

The certified value should be an accurate estimate of the true value with a reliable estimate of the uncertainty. The ISO Guide 35-1985 gives several technically valid approaches for certifying a reference material. Depending on the type of CRM and parameter to be certified, there may be some differences in the approach applied.

One method of certifying a parameter in an RM employs a so-called "definitive method" (e. g. isotope dilution mass spectrometry in inorganic trace analysis) in one single laboratory. However, as said before, this laboratory may have a bias (in fact, it has been often observed that results of definitive methods can be biased because of operator insufficiencies or other reasons) and therefore the certified value may by wrong. This method of using results of one method in one single laboratory may not give a fair estimate of the uncertainty either. In addition, such "definitive methods" hardly exist in organic trace analysis and may present difficulties in some inorganic analyses (e.g. for Hg determination by Isotope Dilution Inductively Coupled Plasma Mass Spectrometry in fish tissue in which high amounts of methyl-mercury may interfere at the ionisation step, [11]. The approach more often adopted consists of using several independent methods including the definitive ones where possible. Certification becomes possible when sufficient evidence is available that two or more independent methods as applied in different high quality laboratories give the same results.

Such laboratories should work under a strict quality assurance regime. It is clear that a basic requirement for the laboratories to avoid systematic errors is that a proper calibration is performed on all the apparatus used. Calibrants must be of adequate purity and of known stoichiometry (special precautions must be taken when all laboratories use the same calibrant if only one supplier exists) and pure solvents and reagents must be used. Chemical reaction yields should be known accurately, and precautions taken to avoid losses (e.g. formation of insoluble or volatile compounds, incomplete extraction) and contamination. If results of entirely different methods of analysis (comparison of methods) applied in different laboratories working independently (comparison of laboratories) are in agreement, it can be concluded that the bias of each method is negligible and the mean value of the results is the best approximation of the true value.

8.4
The Use of RMs and CRMs in Chemical Analysis

The use of CRMs has been described in detail e.g. by Taylor [12] and Griepink and Stoeppler [1]. Some basic principles are described below.

8.4.1
The Role of Reference Materials

RMS can be:

a) pure solutions to be used for testing methods e.g. for inorganic analyses;
b) materials intended to test step(s) of an analytical procedure (e.g. in organic or speciation analysis: pure extract or solution for testing the separation step, raw extract for testing the clean-up procedure, spiked sample for the assessment of extraction recovery);
c) laboratory reference materials having a composition which is as close as feasible to the matrix (including minor and trace constituents of interest) to be analyzed by the user and which have a known content (target value); these materials may be produced in the laboratory to verify the long term reproducibility of analyses performed routinely and may be used for the evaluation of the performance of laboratories in intercomparisons.

8.4.1.1
The Use of RMs in Statistical Control Schemes

Reference materials may be used for the verification of the long term reproducibility of a method by setting up control charts [1, 13]. As soon as a method is considered to be under statistical control i.e. fluctuations in results are only random and not systematic, control charts should be made to detect possible systematic fluctuations (e.g. drift). A control chart is a graphical representation of the results obtained on a reference material in time. One RM should be analysed with 10–20 unknown samples, depending on fluctuation frequency. In order to be able to detect non-random fluctuations in the real analysis, the RM used should pose the same or similar problems to the analytical chemist as the unknown samples do (analytical similarity) and their composition should be homogeneous and stable in time.

In a Shewart control chart (Figs. 1 and 2) the laboratory plots for each RM analyzed at a time the obtained value ($\bar{X}$) or the difference between duplicate values (R). The $\bar{X}$-chart presents the lines beyond which a fluctuation has the probability of 95 or 99% of being of systematic origin. These lines are for "warning" and "action" respectively (Fig. 1). The results of a method are considered to be out of control if (1) the upper or lower control limit is exceeded ("action"), (2) the same "warning line" is exceeded twice in succession and (3) more than ten successive measurements are on the same side of the target value line ("means"-line).

For a more rapid detection of systematic errors, cusum (cumulative sum) charts can be used (Fig. 3). In these charts the cumulative sum of the differences between found values and probable values are plotted against the result

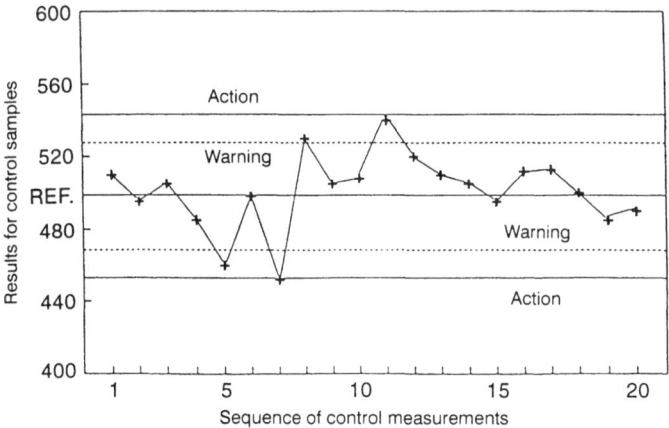

Fig. 1. Example of $\bar{X}$-chart (arbitrary units). $\bar{X}$ is the value obtained on each occasion of analysis. Warning and action lines correspond to a risk of 5 or 1% that the result do not belong to the whole population of results (or in other words that a fluctuation has the probability of 95 or 99% of being of systematic origin

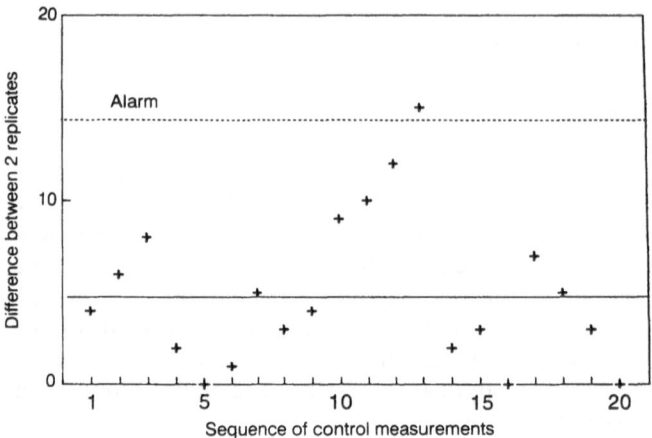

Fig. 2. Example of R-chart (arbitrary units). R is the difference between two duplicate determinations

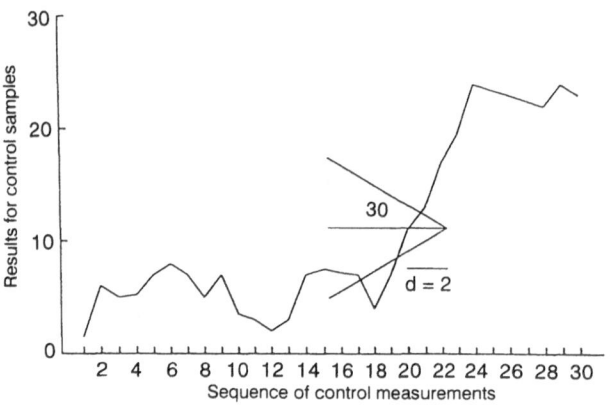

Fig. 3. Example of cusum chart (arbitrary units). The sum of the differences between the result and a reference value are plotted in time. They allow the detection of drifts in methods and trends (d and 30°: V-mask characteristics defined by the user, see Hartley (1990))

obtained at moment i ($S_i = \Sigma \bar{X}_i - \bar{X}_{ref.}$) in which S_i is the cumulative sum, $\bar{X}_i$ is the measured value and $\bar{X}_{ref.}$ is the reference value. As differences accumulate, $\bar{X}_{ref.}$ should be accurately known; therefore, CRMs are preferably used in this case.

8.4.1.2
Use of RMs in Intercomparisons

RMs may be used to assess the performance of analytical techniques in intercomparisons [9]. They can be solutions or extracts to test part of analytical methods (e.g. detection, separation) or samples representative for the type of analysis to be carried out for testing the whole procedures. RMs for intercomparisons are distributed to the participants by a central laboratory; these are requested to perform a given number of replicate determinations of a specified analyte(s). The results may be discussed in technical meetings where the sources of error are identified and may be later removed in the laboratory. By comparing different techniques with different analytical steps (e.g. different extraction, separation, derivatization, detection), sources of error can be detected (e.g. bias due to one particular technique). Similarly a laboratory's performance with a given technique can be evaluated by comparing the results obtained by this laboratory with those of other laboratories using the same technique. As above, it is paramount that the RMs used in such studies are homogeneous and stable.

The discrepancies detected in interlaboratory exercises are thoroughly discussed in various organizations which, in turn, allows a better quality control of the analyses concerned to be achieved. Typical examples found in intercomparisons organized by the BCR illustrate this purpose [14]: some examples have shown that even if a majority of laboratories are in agreement, the results can be wrong. Statistics detect outliers; these can be the accurate values, however, Such errors are not rare when techniques with different analytical steps are involved. This was demonstrated by an intercomparison on hexachlorodibenzonfuran in which only one laboratory, on a group of 10, could demonstrate a complete separation of two compounds (F 118 and F 119) whereas the other laboratories could not separate these compounds using their procedures and reported a value for the compound F 118 only (actually corresponding to the sum F 118 – F 119); the single laboratory was identified as an outlier but was the only one, however, which produced accurate data. Another example of discrepancy was suspected to be due to an incomplete digestion of organic mercury using microwave procedures for the analysis of total mercury in fish [15]; further investigations have shown, however, that the problems were rather linked to ionization problems in the ICP-MS detection. These examples show that whatever the efforts made in improving the quality of chemical analysis and the skill of experience attained, extreme care has still to be taken in ensuring a good quality assurance which can never become a standard routine practice.

Intercomparisons may also be considered as a good tool in the design of a certification campaign as these exercises may allow one to (i) test the preparation of a candidate reference material (homogeneity, stability), (ii) detect and remove sources of systematic errors and (ii) select a group of expert laboratories. These exercises are useful for establishing a check list that should be used by the participants to achieve a good quality control of the analyses; these concern the precautions to be undertaken to avoid systematic errors leading to losses or contamination at various steps of the analytical procedures, e.g. preparation (e.g. errors in weighing, diluting etc). extraction/digestion (e.g. risks of adsorption and desorption, contamination by reagent or vials etc.), derivatization or separation (e.g. incomplete recovery), detection (e.g. interferences) and calibration (e.g. calibrants not sufficiently pure, stoichiometry not verified).

8.4.2
The Role of Certified Reference Materials

CRMs are also referred to as "Standard Reference Materials" by some organizations (e.g. NIST, USA). The use of CRMs is one of the ways to accuracy. Reference materials certified according to the rules laid down in ISO-guide 35 render the user's results traceable to those of the international scientific community. Additionally, they enable the user to verify his performance at any desired moment. CRMs can be defined according to the three following main categories:

a) pure substances or solutions to be used for calibration and/or identification;
b) matrix reference materials which, as far as possible, represent the matrix being analyzed by the user and which have a certified content (such materials are mainly intended for the verification of a measurement process). These materials can be used for the calibration of a certain type of measuring instrument (e.g. spark source emission spectrometry, XRF and those techniques which require a calibration with a material similar to the matrix analyzed);
c) methodologically-defined reference materials for parameters such as: leachable or aqua-regia soluble fractions of trace elements from soils, ashes and slags; bioavailability of a certain element; chloroform-extractable pesticides, etc.; the certified value is defined by the applied method following a very strict analytical protocol.

CRMS are products of very high added value. Their production and certification is very costly (typically some hundred thousands of ECUs) and therefore they should be reserved for final verification of analytical procedures only and not for routine analysis or for external QA (round-robins). CRMs are generally produced in large quantities; they are used for the purpose given below:

8.4.2.1
Calibration

CRMs of pure compounds may be used for calibration (e.g. pure PAHs produced by the US Environmental Protection Agency). Matrix CRMs should not be used for the purpose of calibration unless suitable calibrants are not available. In certain cases, however, some methods (e.g. spark source mass spectrometry, wavelength dispersive X-ray fluorescence, etc.) have to be calibrated with matrix CRMs of a similar, fully characterized matrix (e.g. metal alloys, cements). For such methods, accuracy can only be achieved when certified reference materials are used for the calibration.

8.4.2.2
Achieving Accuracy

When developing a new analytical method or apparatus and after having evaluated all its critical points the analyst has to prove the accuracy of the results obtained. Usually the results will be compared with those obtained with a classical method. This implies that the classical method is fully under control in the laboratory. However, some critical factors such as the influence of laboratory contamination cannot be solved in the laboratory itself and have to be investigated externally by participating in appropriate interlaboratory studies. This allows us to compare the laboratory results with those of other, preferably outstanding, laboratories. A CRM does so too, but whenever necessary and instantaneously [1, 2], CRMs may also be used to test a standardized method when it is applied for the first time in a laboratory.

In most chemical analyses, the measurement process contains steps where the sample is physically destroyed (e. g. acid digestions, fusions, dry ashing) or the analyte to be determined is extracted from the matrix. To ensure accuracy it is necessary to demonstrate that no losses or contaminations occur during such sample treatment. The entire analytical procedure may be verified by using a CRM with a matrix similar to the unknown sample. Disagreement between the certified value and the value determined by the laboratory indicates an error, or errors, in the analytical procedure.

8.4.2.3
Other uses of CRMs

CRMs can also serve the purpose of demonstrating the equivalence of methods, enabling laboratories to follow the developments of new analytical instrumentation. The analyst can compare the performance of his method with those from other laboratories without the need for intercomparisons. Such a use is essential when harmonization of measurements is necessary at an international level.

8.4.3
Suppliers

There are a number of suppliers of RMs and CRMs for the various fields of analyses. Some of them are specialized in a particular field of interest (e.g. National Research Council of Canada, NRCC specialized in marine analysis [7]). Two main bodies, the National Institute of Science and Technology, NIST (USA) and the Community Bureau of Reference (BCR, Commission of the European Communities, Belgium [16]) cover several fields and ensure long term availability of CRMs due to the large batches of materials produced. The International Atomic Energy Agency in Vienna mainly provides materials for nuclear measurements but also supplies some RMs for non-nuclear analysis [17]. The ISO Council on Reference Materials REMCO [18]) publishes a directory for Reference Materials. IUPAC issues a catalogue of CRMs that are available [19]. Some additional compilations of existing CRMs in more specialized fields, e.g. marine monitoring, also exist (e.g. NOAA Data base [4]). The major source of information on reference materials, however, is the COMAR Data Bank which is a joint enterprise between the Laboratoire National d'Essais (Paris, France), the Bundesanstalt für Materialforschung und Prüfung (Berlin, Germany) and National Physical Laboratory (Teddington, Great Britain).

8.4.4
CRMs for Environmental Analysis

The CRMs for environmental monitoring are mainly materials certified for chemical (Griepink et al., [20]), biochemical, microbiological (Mooijman et al., [21]) or sometimes physical parameters.

Classical matrices such as sediments and water are analyzed for environmental monitoring of series of elements and compounds (e.g., major compounds, heavy metals, chemical species. PAHs, pesticides, PCBs, dioxins etc.). CRMs are a necessary tool for the final validation of the methods used in monitoring campaigns. The existing CRMs are mainly river, lake, estuarine and marine sediments, as well as fresh-, ground- and seawater certified for trace elements or trace organics.

Animal tissues or plants are also used in environmental analysis as they often serve to evaluate the global contamination of the environment (target animals or plants) or because they are representative for food chain contamination; the CRMs produced so far for this purpose include mussel, oyster, shrimp, plankton and fish tissue as well as aquatic plants (e.g. sea lettuce) and terrestrial plants (e.g. rye grass, hay powder).

Several industrial products e.g. coals may also serve for the economical evaluation of materials as well as for testing potential contamination by some toxic compounds or elements. Therefore, coal CRMs may be of interest for environmental monitoring.

8.4.5
CRMs for Food Analysis

Measurements for food and agriculture play a key role in several areas, namely for human and animal nutrition, control of undesirable substances in food and feed and, for determining the commercial value of agro-food products [22, 23]. Matrices for human nutrition e.g. meats, cereals and milks are usually analysed for their contents in toxic organic (e.g. PAHs, PCBs) or inorganic substances (heavy metals, major elements), nitrate and pesticides as well as for their nutritional quality; other substances such as e.g. shellfish toxins, mycotoxins and heterocyclic amines are also often considered. Nutritional properties (e. g. fatty acids, sterols, total fat, proteins, carbohydrates oils and fat soluble vitamins etc.) are also being considered in matrices such as vegetables, oil, milk, animal fat, muscle and liver tissue, flour, bread etc. Finally, veterinary drugs are analysed in animal urine, muscle and liver as well as in eggs.

It is clear that there is an increasing need of production of CRMs in the field of food and agriculture analysis in order to achieve good quality control.

8.4.6
CRMs for Clinical Analysis

Accurate determinations in the field of clinical chemistry are particularly important; they indeed serve as a basis for the adoption of reference methods which are being used by a great number of organizations [24]. Serum and blood (in lyophilized form) RMs are most often used for the quality control of e.g. toxic elements (e.g. Cd, Pb) or main group elements (e.g. Ca, Li, Mg) and hormones (e.g. cortisol, progesterone).

Partially or highly purified proteins are also used for the calibration of e.g. thromboplastin used in the determination of the coagulation time of plasma, and apoliproteins. Finally, partially or highly purified enzymes are used to help the standardization of results of measurements of enzyme catalytic concentrations; the enzymes of most concern are e.g. γ-glutamyltransferase, alkaline phosphatase, creatine kinase BB, alanine aminotransferase etc.

Finally, manufacturers of analytical apparatus also produce RMs of serum for controlling the quality of measurements of biochemical parameters. However, in most of the cases these RMs are not comparable from one manufacturer to another which highlights the need for CRM production.

8.4.7
Other CRMs

Industrial products are also being used to provide means of calibration of automatic analyzers. Series of CRMs have been produced for this purpose including non-ferrous metals (e.g. aluminium, copper, lead, nickel, titanium, zirconium etc.) certified for their content of non-metals (e.g. O, C, B, N) or non-metals and alloys (e.g. Cu, Pb, Zn) certified for their trace element content. These CRMs are mainly produced by international corporations such as the Steel and Cement industries.

8.5
References

1. Griepink B, Stoeppler M (1992) Quality assurance and validation of results. In. Hazardous Metals in the Environment. Stoeppler M (ed), Elsevier 17:517–534
2. Maier EA (1991) Certified reference materials for the quality control of measurements in environmental monitoring. Trends in Analytical Chemistry 10:340–347
3. Griepink B (1990) The role of CRMs in measurement systems. Fresenius' J Anal Chem 338:360–362
4. ISO (1985) Certification of reference materials – General and statistical principles. ISO/IEC Guide 35-1985. International Organization for Standardization, Geneva, Switzerland
5. Griepink B, Muntau H (1988) EUR Report. 1185 EN, European Commission, Brussels

6. Quevauviller Ph, Kramer GN, Griepink B (1992) A new certified reference material for the quality control of trace elements in marine monitoring: cod muscle (CRM 422). Mar Pollut Bull 24(12):601–606
7. NRCC (1990) BCSS-1, MESS-1, PACS-1, BEST-1. Marine sediment reference materials for trace metals and other constituents. National Research Council Canada, Ottawa
8. Allen DW, Brookst JS, Unwin J, McGuiness D (1989) Studies on the degradation of organotin stabilizers in poly(vinyl chloride) during gamma irradiation. Appl Organometal Chem 1:311–317
9. Maier EA, Quevauviller Ph, Griepink B (1993) Interlaboratory tudies as a tool for many purposes: proficiency testing, learning exercises, quality control and certification of materials. Anal Chim Acta 283:590–599
10. Quevauviller Ph, Maier EA, Vercoutere K, Muntau H, Griepink B (1993) Certified reference material (CRM 397) for the quality control of trace element analysis of human hair, Fresenius' J Anal Chem 343:335–338
11. Campbell MJ, Vermeir G, Dams R, Quevauviller Ph (1992) Influence of chemical species on the determination of mercury in a biological matrix (cod muscle) using inductively coupled plasma mass spectrometry. J Anal At Spectr 7:617–621
12. Taylor JK (1993) Handbook for SRM-Users (NBS Special Publ), No 260–100, pp 100, 2nd edition
13. Hartley TH (1990) Computerized Quality Control: Programs for the Analytical Laboratory. Ellis Horwood, Chichester. 2nd Edition, pp 99
14. Griepink B, Quevauviller Ph, Maier EA, Vandendriessche S (1993) BCR: a service to quality assurance in analytical chemistry – some experiences and achievements with regard to reference material preparation. Fresenius' J Anal Chem 346:530–535
15. Quevauviller Ph, Imbert IL, Ollé M (1993) Evaluation of the use of microwave oven system for the digestion of environmental samples. Microchim Acta 112:147–154
16. BCR Catalogue (1992) BCR Reference Materials Community Bureau of Reference (BCR). Commission of the European Communities. Rue de la Loi 200, 1049 Brussels. Belgium
17. Cortes Toro E, Parr RM, Clements SA (1990) Biological and environmental reference materials for trace elements, nuclides and organic microcontaminants – A survey. IAEA/ RL/128 (Rev 1). International Atomic Energy Agency, Vienna
18. Directory of Certified Reference Materials. Secretary for REMCO, ISO. Case Postale 56, 1211 Geneva, Switzerland
19. Cantillo AY (1992) Standard and reference materials for marine science. NOAA National Status and trends program for Marine Environmental Quality. US Department of Commerce. Rockville, MD, USA, third edition, pp 575
20. Griepink B, Maier EA, Quevauviller Ph, Muntau H (1991) Certified reference materials for the quality control of analysis in the environment. Fresenius' J Anal Chem 339:599–603
21. Mooijman KA, In't Veld PH, Hoekstra JA, Heisterkamp SH, Havelaar AH, Notermans SHW, Roberts D, Griepink B, Maier EA (1992) Development of microbiological reference materials. EUR Report 14375 EN. CEC Brussels, pp 122
22. Belliardo JJ, Wagstaffe PJ (1988) BCR reference materials for food and agricultural analysis: an overview. Fresenius' J Anal Chem 332:533–538
23. Cornelis R (1992) Use of reference materials in trace element analysis of foodstuffs. Food Chem 43:307–313
24. Colinet E (1992) Quality assurance in the field of biomedical analyses: the role and the contribution of the BCR Program of the European Communities. Microchem J 45:237–240

Chapter 9

Accreditation and Interlaboratory Studies

W. P. Cofino

9.1
Introduction

The quality of products, services and commodities is an important issue in modern society. It is generally realized that poor quality can cause economic losses or impair human health or the environment. Customers increasingly demand evidence of the quality of products or services. Accreditation meets this need – it serves as a token that an organization supplies what it lays claim to.

Organizations that pursue accreditation have to implement a quality system according to a specific standard. For testing and calibration laboratories, accreditation is based on EN 45001 [1] or ISO 25 [2]. These standards contain criteria which administrative and technical procedures have to comply with. The criteria include practices to verify the quality of testing, among others through participation in interlaboratory studies.

This paper discusses interlaboratory studies in accreditation. The different types of interlaboratory studies are outlined in Sect. 9.2. Section 9.3 considers the manner accreditation bodies deal in practice with the stipulations in EN 45001 related to interlaboratory studies. In paragraph four, the importance of interlaboratory studies for accreditation is discussed in a broader context.

9.2
Types of Interlaboratory Studies

Three major types of interlaboratory studies can be discerned, respectively focussing on the performance of methods, the performance of laboratories and on the assignment of a most probable value of a quantity to a material with a stated uncertainty. A nomenclature for interlaboratory analytical studies is in preparation and provides the following operational definitions [3]:

- An interlaboratory is a study, in which several laboratories measure a quantity in one or more identical portions of homogeneous materials under documented conditions, the results of which are compiled into a single report.
- A method-performance study is an interlaboratory study in which all laboratories follow the same written protocol and use the same test method to measure a quantity in sets of identical test samples. The reported results are

used to estimate the performance characteristics of the method; usually these characteristics are within-laboratory and between laboratory precision, and when necessary and possible, other pertinent characteristics such as systematic error, internal quality control parameters, sensitivity, limit of determination and applicability.

- A laboratory-performance study is an interlaboratory study that consists of one or more analyses or measurements by a group of laboratories on one or more homogeneous test samples by the method selected or used by each laboratory. The reported results are compared with those from other laboratories or with the known or assigned reference value, usually with the objective of evaluating or improving laboratory performance.
- A material-certification study is an interlaboratory study that assigns a reference value ("true value") to a quantity (concentration or property) in the test material, usually with a stated uncertainty.

In the literature, the term proficiency testing is quite often employed. ISO 43 [4] and EN 45001 [1] describe proficiency testing as 'methods of checking laboratory performance by means of interlaboratory tests'. ISO 25 provides a slightly different definition: 'the determination of the laboratory calibration or testing performance by means of interlaboratory comparisons' [2]. A recent IUPAC-ISO/REMCO protocol introduces a proficiency testing scheme, described as "a method of checking laboratory testing performance by means of interlaboratory tests. It includes comparison of a laboratory's results at intervals with those of other laboratories, with the main object being the establishment of trueness" [5]. This paper adheres to the proposed IUPAC definitions and does not use the terms proficiency tests or proficiency testing schemes. Instead, reference is made to laboratory-performance studies and laboratory-performance study schemes [3].

All three types of interlaboratory studies are important to achieve quality of testing. This is acknowledged in ISO 25 and specifically in EN 45001. The latter standard contains a paragraph on the cooperation with other laboratories and with bodies producing standards and regulations [6]. Implicitly. EN 45001 concedes the need for well-tested, validated methodology and certified reference materials and encourages the organization of and participation in method-performance and material-certification studies. ISO 25 and EN 45001 pay more attention to laboratory-performance studies. The remainder of this paper will be confined to this type of interlaboratory study.

9.3
Laboratory-Performance Studies in Accredition Practice

Accreditation systems use criteria that are based on EN 45001 and ISO 25. Both standards contain clauses about interlaboratory studies. The Western European Laboratory Accreditation Cooperation WELAC recently has issued criteria for proficiency testing in accreditation [7]. The clauses in ISO 25 and EN 45001 and the WELAC criteria will be discussed in this section.

9.3.1
Objectives of Participation in Laboratory-Performance Studies

Laboratory-performance studies provide feedback on three levels [8]:

1) Individual laboratories obtain information on their performance and can objectively assess and demonstrate the reliability of their results [5].
   ISO 25 and EN 45001 define two objectives for participation in laboratory-performance studies:

   - Wherever applicable, measurements ought to be traceable to national or international standards. A laboratory has to provide evidence for accuracy when traceability cannot be achieved. Participation in a suitable interlaboratory study is indicated as a possibility [9, 10].
   - the quality of testing has to be ensured by among others, by implementing checks, one of these being participation in laboratory-performance studies [11, 12]

2) Third parties as accreditation bodies or customers can request a laboratory to provide information on their performance.
   EN 45001 states explicitly that an accredited laboratory has to permit the accreditation body to scrutinize the results obtained in laboratory-performance studies.
3) The whole sector is informed on the state of the practice with respect to analytical proficiency. Poor overall performance points to the need for harmonization of methodology, validation of harmonized methods via method-performance interlaboratory studies, and/or the preparation of (certified) reference materials. Poor overall performance may also point to inadequate design of for instance monitoring programmes owing to insufficient specification of analytical methodology.

ISO 25 states that a laboratory must participate in interlaboratory studies where appropriate [13]. EN 45001 makes a similar statement [14]. EN 45001 pronounces in addition, that a laboratory must participate in any appropriate programme of interlaboratory studies which the accreditation body reasonably seems necessary, [15]. ISO 25 and EN 45001 both state that the quality manual should make reference to verification practices including interlaboratory studies.

The WELAC criteria follow the stipulations put forward in EN 45001. WELAC recognizes the problem that mandatory participation in laboratory-performance studies for all parameters would be a large burden for laboratories which provide a wide range of analytical services. The WELAC criteria imply, that a laboratory should develop a policy for participation in interlaboratory studies. This policy may entail, for instance, that different methods are assessed alternately through laboratory-performance studies. The policy has to be approved by the accreditation body. Both individual laboratories or groups of laboratories may reach an agreement with the accreditation body [7].

9.3.2
Assessment of Laboratory Performance

Laboratory-performance studies are realized in a variety of ways and are not carried out according to the same level of quality. A laboratory has to make its own evaluation as to whether the organization and realization of the inter-laboratory study justifies the drawing of conclusions with respect to its performance. Factors which a laboratory (and an accreditation body) have to take into consideration, in particular when comparing results to those of other laboratories, are [16, 17]:

- the origin and character (e.g. matrix characteristics, concentration levels) of testing samples;
- the number of participants and their experience with the methodology tested;
- the test methods used and, where possible, the assignment of the results to particular methods;
- the organization of the study (e.g. the statistical model, the number of replicas, the parameter to be measured, the method for establishing the assigned value);
- the criteria used by the organizing body to evaluate the participants' performance.

Central in the assessment of laboratory performance is a consideration of bias. Bias is estimated as the difference between the result of the laboratory and the assigned value, the latter being the best estimate of true value. A performance criterion is required to judge the magnitude of the bias. To this end, it is possible to define domains centred around the assigned value to which certain performance-levels are attached. Frequently, a certain between-laboratory standard deviation is chosen to establish these domains on statistical grounds. In particular, accreditation bodies may appraise classification of laboratory performance. ISO/REMCO N280 states, however, that classification is not recommended in laboratory-performance studies and suggests the use of 'decision limits' based on z-scores as an alternative [5].

Z-scores entail a transformation of laboratory results according to the formula:

$$z_i = \frac{x_i - X}{\sigma}$$

where z_i: the z-score for laboratory i

x_i: the result of the laboratory i

X: the assigned value

σ: the (between-laboratory) standard deviation chosen as performance criterion.

The z-score renders the estimated bias of a laboratory in units of the standards deviation which has been chosen as performance criterion. It is possible to classify the z-scores (or to establish decision limits using z-scores) when a Gaussian distribution with the assigned value as mean and the chosen standard deviation is assumed valid or is imposed as a model to evaluate the data. For such a normal distribution, $|z|$ would be two or less in 95% of the cases while the probability that $|z| \geq 3$ is about 0.3%. Therefore, a reasonable classification would be that z-scores are satisfactorily when $|z| \leq 2$, questionable when $2 < |z| < 3$, and unsatisfactorily when $|z| > 3$ [5]. It is noted that a certain degree of uncertainty will be associated with the assigned value. Assessments of performance need to take this uncertainty into account.

It is WELAC policy to adhere as much as possible to international standards and will therefore follow the guidelines delineated in standards such as ISO/REMCO N 280 [5] as the basis for the assessment of the performance of laboratories on the basis of laboratory-performance studies. WELAC has extended the guidelines put forward in ISO/REMCO N 280 to some extent, also adopting a Gaussian distribution as model. The WELAC guidelines, expressed in tems of z-scores, entail that the performance of a laboratory for a particular determinand is considered to be unsatisfactory when

1) a result is an outlier or has an absolute z-score greater than three;
2) the results obtained for a number of samples in one laboratory-performance study have absolute z-scores greater than two;
3) the results for a particular parameter have absolute z-scores greater than two in successive laboratory-performance studies of a similar nature.

WELAC does not explicitly state how to arrive at the assigned value and indicates that the standard deviation derived from the between-laboratory variance (reproducibility) can be used to calculate the z-score. WELAC emphasizes, however, that their guidelines are not exhaustive but are provided to illustrate the line reasoning [7].

Soundly derived assigned values and standard deviations chosen as performance criteria are essential. Two main lines may be followed:

– the average and standard deviations of the laboratory-performance study computed after statistical treatment of the data are employed;
– the assigned value and (between-laboratory) standard deviation are established independently.

Statistical treatment of laboratory-performance studies often invokes the application of outlier tests (e.g. Cochran, Grubbs, Dixon) followed by computation of the average and standard deviation, comparable to the procedure outlined in ISO 5725 [18]. Such treatments proceed from the assumption that the data follow a Gaussian distribution and that the variances of all laboratories are equal. Such assumptions are frequently violated in practice. Presently, the use of robust statistics is advocated [19–21]. This type of statistics does not rely on a particular distribution of the data. Reports on the application of robust statistics are promising [22, 23].

The average and standard deviations calculated by different statistical methods may differ. More importantly, these parameters describe the population of data of the participating laboratories and are as such subjective. The proximity of the average of the true concentration and the magnitude of the standard deviation will depend *inter alia* upon the general experience with the methods, the concentration level and matrix tested, the types of methodology employed, the degree to which methodology has been validated and harmonized, the availability of certified reference materials for the particular determinand and matrix and the proportion of qualified laboratories in the study. The assumption that the average reflects adequately the true value is not assured by a large number of participants nor be the presence of a group of laboratories reporting similar results. Great caution should be taken when taking the consensus concentration and standard deviation obtained from the laboratory-performance study itself for the assessment of laboratory performance. This approach may be acceptable for simple, well crystallized methods or for empirical methods where the analyte is operationally defined. In general, however, it is considered less appropriate [5].

Approaches for the independent assessment of the assigned value and the (between-laboratory) 'target' standard deviation are described in ISO/REMCO N 280 [5].

The assigned value may to obtained as [5]:

- the consensus value from a group of expert laboratories;
- the addition of a known amount or concentration of analyte to a base material containing none;
- direct comparison of the test material with certified reference materials by analysis with a suitable analysis under repeatability conditions.

A value for the target (between-laboratory) standard deviation can be obtained along a number of lines [5]:

- by estimation of the between-laboratory precision required for the intended purpose of the measurement;
- by perceiving how laboratories should perform;
- by reference to the reproducibility obtained during appropriate method-performance studies;
- by reference to a generalized model, e. g. the 'Horwitz curve' [24].

ISO/REMCO and WELAC emphasize, that the performance criteria ought to be related to the performance characteristics required for the objective of measurement and not be based on the state of the art [5, 7]. In all cases, options and results have to be scrutinized carefully. The organizers need to have or invoke a high level of experience and expertise with the methods and matrices tested and with statistics.

The WELAC criteria state, that a laboratory has to record the outcome of the assessment of its own performance for each laboratory-performance study in which it has participated. If necessary, corrective actions have to be taken and documented. In addition, when a laboratory does not take part in a laboratory-performance study which was made mandatory by the accreditation body, it has to note the reasons for failing to do so [7].

An accreditation body may withdraw the accreditation for a particular test or kind of tests when a laboratory does not take corrective actions or when these actions appear to be not effective, as judged by the results in successive laboratory-performance tests. The accreditation body will assess a laboratory thoroughly, considering the entire quality control system, before taking such a measure [7].

9.3.3
The Implementation of Laboratory-Performance Studies

The results of a interlaboratory study provide important information to both laboratory and accreditation bodies. The quality of the interlaboratory study itself is therefore an issue which has to be addressed. The WELAC document provides criteria for the design and execution of laboratory-performance studies [7]. The criteria constitute, with the stipulations laid down in EN 45001 or ISO 25, the basis for a quality system for interlaboratory studies. The organization has to have appropriate managerial and technical (methodology, statistics) qualifications. Certa n gaps in technical expertise (e.g. statistics) may be accounted for by invoking external experts. A qualified study manager has to be appointed. Arrangements for quality audits have to be made. A complete study file has to be prepared and maintained. The design of the laboratory-performance study has to be thoroughly documented in a study plan. The laboratory-performance study file has to include all correspondence with the participants. Particular attention has to be paid to communication with the participants. The latter have to be informed completely about, for instance, the objectives of the laboratory-performance study, criteria for the acceptance of participants, the activities to be carried out and specific requirements which have to be met, the way data are treated and presented, the degree to which results are made public (including e.g. laboratories coded or not) and the manner in which the results are used. The samples have to be as much as representative as possible for those encountered in daily practice. Homogeneity and stability have to be assured.

The American ASTM is also preparing a standard guide for 'the development and operation of laboratory proficiency testing programmes' [25].

Some accreditation bodies intend to accredit organizers of laboratory-performance studies when they have established quality systems which meet the WELAC criteria.

9.4
Laboratory-Performance Studies and Quality of Testing

The quality of testing in general is of concern. Laboratory-performance studies should form part of a system which assures in general terms the quality of testing. The basic design of such a system is schematically outlined in Fig. 1, which is based on the "Plan, Do, Check, Action" cycle of Deming.

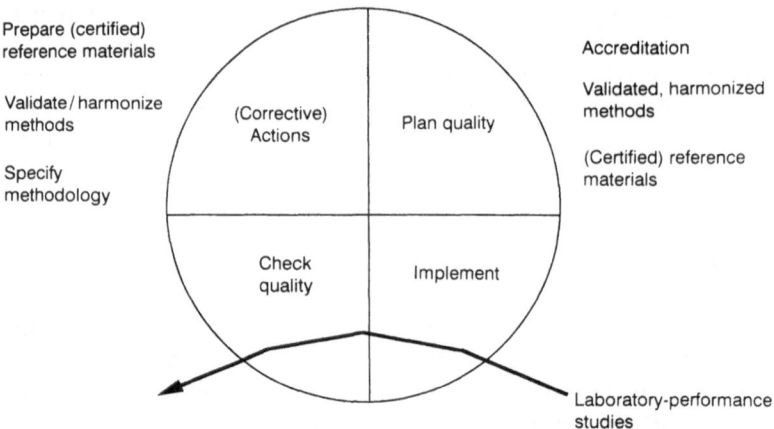

Fig. 1. Outline of system to improve and maintain general quality of testing

Quality of testing has to be planned – measures have to be taken in order to prevent poor quality (as manifested by poor between-laboratory comparability). These measures entail the implementation of (accredited) quality systems, harmonization and validation of (standard) methodology and the provision of (certified) reference materials. In the second phase, measurements are carried out by the laboratories involved. As part of the system under consideration, laboratory-performance studies are organized. Apart from providing individual laboratories with objective feed-back about their performance, information is gained on the overall performance (e.g. see Sect. 9.3 [8]). In the third phase, the results of the laboratory-performance studies are examined, now specifically with respect to the general performance. When unacceptable overall performance is observed, actions should be undertaken (phase 4): merely repeating the laboratory-performance study will just repeatedly highlight the need for measures to be taken. These actions may include the development and validation of e.g. (standard) methodology or (certified) reference materials or specification of methodology. The latter can be particularly important when empirical methods are used (e.g. organic carbon in environmental analysis, [8]). The completion of these actions may be considered as the start of a new cycle.

The approach described here emphasizes that a structure should be created in which laboratories are provided tools and support in developing, puting into operation and validating methodology and verifying performance. The development of such tools and support should be carried out as a connected action – only then, will the full potential of accreditation practices, validated methodology, the availability of (certified) reference materials and the conduct of laboratory-performance studies be utilized.

9.5
References

1. CEN/CENELEC (1989) European Standard EN 45001, 'General Criteria for the operation of testing laboratories', Brussels, Belgium
2. International Organization for Standardization (1990) ISO/IEC Guide 25, 'General requirements for the technical competence of testing laboratories', Third Edition, Geneva
3. Horwitz W (1991) Project 27/87 Nomenclature for Interlaboratory Studies, Fourth Draft, IUPAC
4. International Organization for Standardization (1984) ISO/IEC Guide 43, "Development and Operation of Laboratory Proficiency Testing", Geneva
5. International Organization for Standardization (1993) ISO/REMCO N 280, 'Proficiency Testing of Chemical Analytical Laboratories', Geneva. Republication of the Technical Report "The International Harmonized protocol for the Proficiency Testing of (Chemical) Analytical Laboratories, IUPAC Pure & Appl Chem 65 2123 – 2144
6. EN 45001, section 6.3
7. WELAC Western European Laboratory Accreditation Cooperation, Working Group 4: Proficiency Testing, Revised draft proposal, revision of October 1992
8. Cofino WP (1993) 'Quality Asuurance in Environmental Analysis', in Barcelo D (ed), 'Techniques in Environmental Analysis', Elsevier, Amsterdam
9. ISO 25, clause 9.3
10. EN 45001, clause 5.3.3
11. ISO 25, clause 5.6.c
12. EN 45001, clause 6.3
13. ISO 25, clause 4.2.j
14. EN 45001, clause 6.3
15. EN 45001, clause 6.2.d
16. National Accreditation Board of the Netherlands STERLAB (1991) 'Accreditation and Interlaboratory Studies', STERLAB, Rotterdam
17. Western European Laboratory Accreditation Cooperation (WELAC) (1993) 'WELAC Criteria for Proficiency Testing in Accreditation'
18. International Organization for Standardization (1986) International Standard 5725, 'Precision of test methods – Determination of repeatability and reproducibility for a standard test method by inter-laboratory tests', Geneva
19. Analytical Methods Committee (1989) 'Robust Statistics – How not to reject outliers. Part 1. Basic Concepts', Analyst 114:1693
20. Analytical Methods Committee (1989) 'Robust Statistics – How not to reject outliers. Part 2. Inter-laboratory trials', Analyst 114:1699
21. Lischer P (1990) 'Statistik and Ringuntersuche (Auszug aus dem Schweiz. Lebensmittelbuch)', Schriftenreihe der FAC Liebefeld Nummer 6, Eidgenössische Forschungsanstalt für Agrikulturchemie und Umwelthygiene, Liebefeld-Bern
22. Miller JN (1993) 'Tutorial review: Outliers in experimental data and their treatment', Analyst 118:455
23. Thompson M, Mertens B, Kessler M, Fearn T (1993) 'Efficacy of robust analysis of variance for the interpretation of data from collaborative trials', Analyst 118:235
24. Horwitz W (1982) Anal Chem 54:67A – 76A
25. ASTM Committee E-36, Standard Guide for the development and operation of laboratory proficiency testing programs, in preparation

Chapter 10

Accreditation Competence:
Requirements for Accreditation Bodies

Georg J. Mechelke

10.1
Standard Fundamentals

The standard DIN EN 45003 "was drawn up with the objective of promoting confidence in those accreditation systems and bodies which conform to them and, hence, in testing laboratories assessed or accredited under such systems" (foreword to DIN EN 45003). Confidence in the accreditation competence of accreditation bodies thus determines confidence in the testing competence of the accredited testing laboratories. DIN EN 45002 and DIN EN 45003 therefore impose compulsory duties and obligations to publish, disclose and make statements upon accreditation bodies. The degree of transparency differs for testing laboratories applying and those already accredited and all "interested bodies"; transparency does not arise when confidentiality has to be respected according to clause 13, DIN EN 45003. This standard determines the "General criteria for Laboratory Accreditation Bodies". Adopted by CEN/CENLEC as a European standard on 23.06.1989, having achieved the status of a national standard on 01.05.1990, it will probably be replaced in the near future by the ISO/IEC guideline 58:1993 "Accreditation systems for calibration and testing laboratories – General requirements for operation and acceptance". Thus, the requirements for accreditation bodies will not be presented in an exegetical discussion, but will be dealt with in a more practical and authentic way for both standard and guideline.

10.2
Organization and Quality Management Systems

In order to fulfill its tasks, the accreditation body needs an organization which – independent of the chosen legal status – fulfills the requirements mentioned under clauses 4 and 5 of DIN EN 45003 and sub-clause 4.2 of the guideline. They are characterized by the rules of commercial and financial independance, by impartiality in operation and decision finding processes, by a balanced constitution and competence of the committees, and by the separation of assessment and decision making processes.

It results in an organisational structure of an accreditation body that is basically quite similar in all countries of the European Economic Union. All accreditation bodies who are members of the Deutscher Akkreditierungsrat (German Accreditation Council, DAR) have established committees carrying out the function of "steering committees" and are mostly so called. While the steering committee determines accreditation policy and standard arrangements for accreditation, the accreditation committee decides on whether an accreditation is granted, maintained, extended, suspended, or withdrawn. The complaints committee has to be appealed to in the case of disputes over these decisions.

While respecting the interests of third parties, it is at the discretion of the accreditation body to decide on setting up further committees, on their designation and assignments of duties. Apart from the recommendation of clause 7, EN 45003 and sub-clause 4.2.1k of the guideline to establish one or more sectoral committees, standard and guideline maintain a neutral position on this item.

Neither standard nor guideline specify the personal and professional qualifications which the committee members should have. Since they select and appoint assessors they should be at least as well qualified as the assessors as set out in clause 7, DIN EN 45002 and clause 5 of the guideline.

The quality management system established by the accreditation body has to be appropriately 'tailored' to the type, range and volume of the practised accreditation activities. This rule for quality assuring appropriateness allows the accreditation body great freedom of action in the area of organization. This has to be documented in a quality manual and in quality management operating procedures and has to be checked by internal audits. A third-party auditing is not scheduled in the standard series EN 45000 nor in the guideline. However, it is performed through 'evaluation' by other accreditation bodies in the course of the establishment of international agreements on mutual recognition.

10.3
Arrangements for Accreditation

Arrangements for accreditation constitute the most important part of the accreditation standards of an accreditation body. So to speak these arrangements are the General Conditions of Competence Confirmation of the accreditation body because they define the conditions for the granting, maintenance and renewal of accreditation and the conditions under which accreditation may be suspended or withdrawn. The conditions have to be determined exclusively by competence. Access to the accreditation system described by the accreditation body must neither be conditional upon the size of the laboratory nor upon its membership within any association or group. Often, the arrangements for accreditation or "accreditation guidelines" become an integral part of an accreditation contract signed by applicant and accreditation body. The contractual provisions must strictly distinguish between the accreditation process and the rights and duties of the testing laboratory resulting from accreditation. This avoids the impression of an contractually assured accreditation automatism. The applicant should not take the decision of the accreditation committee as an inevitable occurrence, but as a turning-point between exclusively self-controlled and supervised testing activities. Therefore, on application, it is recommended that a contract for "carrying out an accreditation procedure" be made and an "accreditation contract" be signed after the accreditation committee's positive decision on accreditation. This corresponds to the provision determined by clause 5, DIN EN 45002 and clause 12, DIN EN 45003.

The arrangements for accreditation have to be constantly checked for their appropriateness, practicability, and consistency with the ISO and CEN standards as well as the regulations of TGA, DAR, WELAC, WECC, and ILAC.

The accreditation body is soundly advised to adopt further international developments of the standards as soon as possible in order to promote, through comparability of its standards, the acceptance of test results of testing laboratories to which it has granted accreditation.

10.4
Operation

A differentiation has to be made between the operating procedures specified by the accreditation body's arrangements for accreditation and its method of operation. An accreditation body has to carry out accreditation procedures swiftly and in a non-discriminatory manner. Accreditation procedures should not last more than one year. According to the arrangements for accreditation, an accreditation body has to terminate the procedure if it can only carry it out with interruptions due to insufficient preparation and incomplete information by the testing laboratory. In no case, should an accreditation body enable a testing laboratory – so to say, in the course of an accreditation procedure – to comply with the requirements of DIN EN 45001 and its accreditation criteria step by step. This would contradict the standard which starts from the competence declaration on application which has to be proven according to the arrangements for accreditation. An accreditation body does not occupy the position of a Director of Public Prosecution or Court nor is it a wetnurse or consultant (refer to sub-clause 4.2.1.1 of the guideline). Here, of course, a distinction has to be made with regard to proposing amendments i.e. removals of defects or changes included in the on-site assessment report according to clause 6, DIN EN 45002.

10.5
Sectoral Committees

Sectoral committees play a leading role in the technical operation of accreditation bodies. Clause 7, DIN EN 45003 summarizes in one sentence the tasks of sectoral committees and determines their minimum number. In sub-clause 4.2.1k of the guideline, too, this item is restricted to one sentence. The standard and guideline do not provide information on the members quali-

fications; this is in contrast to sub-clause 7.1, DIN EN 45002 and clause 5 of
the guideline which specify the qualification requirements for assessors. This
is surprising. Without sectoral committees, an accreditation body is unable
to elaborate specific technical criteria deemed as effective requirements
for competence testing in a given field. The sectoral committees determine
the technical quality standard of an accreditation body; assessors check its
achievement and maintenance during the accreditation process and the
subsequent surveillance procedure. Therefore, great demands are made on a
sectoral committee member's technical competence. The members should
have thorough knowledge of and be experienced in quality assurance, be con-
versant with the test methods and types of tests through long years of practical work, and be familiar with up-to-date standardization. It is one of the
special duties of the sectoral committee chairmen to comment on procedure-
related problems arising during onsite assessments. In that way, they may
make on essential contribution to the efficient execution of an accreditation
procedure.

In the frame of the "assessor qualification procedure", sectoral committees
should recommend assessors whose proven competence obviously suggests
that they should be given a basic contract. As a result of the imposed pro-
hibition for self-appointment, sectoral committee members may not be
appointed as assessors.

Supreme importance is given to the exchange of opinions and experience
between sectoral committees in order to guarantee the equivalence of accreditation and assessment requirements and to avoid particularism.

10.6
Assessment

The activity of an accreditation body consists essentially of on-site assessment,
i.e. assessment carried out in the units of the testing laboratory specified in the
application for accreditation. Laboratory facilities not mentioned within the
scope of intended accreditation need not be made accessible to assessors. All
information necessary for preparing the assessment according to subclause
6.2, DIN EN 45002 has to be passed to the accreditation body prior to the
assessment and has to be "evaluated" by the assessors (sub-clause 6.2.1 of the
guideline). Accreditation bodies mostly call this type of evaluation "formal
assessment" according to DIN EN 45001 and 45002. The fulfillment of the
'General criteria for the operation of testing laboratories', expecially the "technical competence" specified by clause 5, has to be assessed on-site. The accreditation body decides on its own whether there are "supplementary technical criteria" to be met by the testing laboratory and which assessment resources
(questionnaire, conduct of test methods) have to be applied when assessing
competence. Of course, consideration has to be given to the textual requirements of the assessment report according to clauses 6 and 9, DIN EN 45002
and to sub-clause 6.4 of the guideline which propose using a questionnaire

on-site. The qualification of the laboratory to perform tests in the contractually specified scope of accreditation which is stated in these reports, presupposes an examination of the practised testing activities. This may be surveyed on-site by assessing the handling of certain test methods. The rule to apply equal assessment criteria does not allow it to be left to the assessor's decision to determine type and number of these test methods. Thus, these test methods have to be determined and carried out according to a system specified by the conditions of the assessment method. Casting lots is an approved method of determining a certain number of test methods relative to the number of the test methods that are related to the individual item of application. Types of tests have to be assessed analoguously. This rules out an assessors's predictable preferences for test methods.

The recording system is reviewed subsequent to the assessment of the actual testing competence. The recording system is checked by an examination of the quality assurance manual, the instrument and/or test method manuals, and test-related documents such as operating procedures and instructions for quality assurance procedures. Apart from test records, special attention has to be drawn to the calibration, maintenance and testing protocols of the manuals because here, a direct insight into realized and not only documented quality assurance is given. Consultation of written documents is considered equal to viewing data on monitors. This constitutes a check for completeness, plausibility and comprehensibility of data stored in a laboratory information system for one special test. Documented quality assurance is then a conclusive proof for a well-functioning record system when all data related to daily testing operations are found and are comprehensible.

The recording system according to DIN EN 45001 does not bind the testing laboratory to respect either a given structure or a terminology. Thus, there are "device manuals", "device volumes", "test method manuals", "test protocols", "calibration and maintenance protocols", "certificates of analysis", "laboratory information systems". In view of the terminology used and understood for years in a laboratory, care has to be taken when assessing the recording system. However, the laboratory is required to provide a tight structure of the system, contextual completeness according to EN 45001, terminological clarity and know-how of the laboratory staff.

A recording system convincing in its structure, content and design is imperfect if the individuals misunderstand either the terms used in the regulation forms or their purpose.

Assessment criteria and the assessment method have to be announced to the testing laboratory on application. It is not within the task of the accreditation body to surprise the laboratory by asking questions, but to reconstruct and comprehensively document the competence declaration of the laboratory. This means, a laboratory aware of the requested requirements showing weak points or even shortcomings contradicts its competence statement on application. Imperfections in the recording system reveal declaratory rather than textual deficiencies. Imperfect handling of test methods under application is always a conlusive proof of incompetence.

10.7
Assessors

The quality of an assessment is governed by the assessor's competence related to test methods and types of tests. However, the accreditation body has to observe, too, whether further requirements of clause 7, DIN 45002 and sub-clause 5.1 of the guideline are fulfilled. Here, assistance is rendered through the "procedure for qualifying assessors" elaborated on the accreditation body's own responsibility. Assessors have to be selected, have to be appointed as assessors of the accreditation body by contractual arrangements, and have to be entrusted for each individual case.

The appointment of an individual assessor or the constitution of an assessment team is governed by the scope of intended accreditation and the avoidance conflicting interests between the laboratory to be assessed and the assessors. Furthermore, when appointing an assessment team, high significance has to be addressed to the expected cooperation between the lead assessor and the other assessors.

Their tasks and competence may not be defined by hierarchical order but along personal and technical lines of team-work. The assessor does not assist the lead assessor but assesses according to his own technical know-how and on his own responsibility. Thus, only well-founded cases justify, that the lead assessor will – and has to – disregard the assessor's assessment report or his opinion. Malevolence excluded, this can only be the case for an incorrect assessments through negligence. This, a know-all manner, or a schoolmasterish attitude blacklists an assessor from all (further) assessment.

During on-site assessment, the assessor may have a considerable motivating or demotivating effect; especially on staff performing the test methods. Thus, he should keep in mind the need to prevent nervousness during demonstrations, he should praise when justified, and pass on constructive criticism. In no case should he act as an examiner giving marks. This would contradict "accreditation" as being a competence certification that does not affect performance and that is free of valuation. Not discussed here are possible constitutional problems in labour-management relationships that may arise for testing laboratories and accreditation bodies. Unsuited for an assessment are also those individuals who want to inaugurate standard modifications by means of an assessment; but, if these demands for reforms constitute the result of an assessment, then, they may promote standardization work and, hence, will help to improve testing activities for "all interested circles".

10.8
Decision on Accreditation

The decision as to whether or not to accredit a testing laboratory has to be taken by the accreditation body on the basis of the result of the assessment (sub-clauses 6.5 and 6.6, DIN EN 45002 and sub-clause 6.5 of the guideline). According to sub-clause 6.6, EN 45002, it is allowed to grant accreditation for a limited period and to attach certain conditions. A time limit is general practice in Europe, except in Sweden. The granted accreditation period is between 2 and 5 years. The possibility given by the standard to grant competence certification under certain conditions gives cause for potential conflicts and is no longer valid according to the guideline. This is welcome. The competence declaration of the laboratory has to be either confirmed or denied. Neither can a competence declaration be restricted by conditions or charges nor can it be qualified by marks or points. Here, one has to differentiate regulations and directions stipulated as prerequistite for granting an accreditation. Accreditation bodies, granting accreditations with attached conditions or charges neither promote comparability of test results nor do they contribute to building up confidence; on the contrary, they foster competitive distortions among testing laboratories and, finally, undermine confidence in a proven product quality.

10.9
Diligence and Protective Duties

Because of their relevance for market competition in the relevant testing field, tasks determined in clause 10, DIN EN 45003 and in sub-clauses 4.5f and 6.6 of the guideline entitled "Accreditation Documents" and "Granting of Accreditation" have to be performed with discretion. Accredited laboratories with confirmed competence enjoy a competitive advantage over laboratories with unproven, or according DIN EN 45001, not even claimed competence. Accredited laboratories are in competition within the same testing fields. Thus, the scope for which accreditation is granted has to be specified precisely in the accreditation documents. This means avoiding effusive terminology on the first page of the accreditation document, and demands that on the following pages, e.g. according to DAR, in the "Annex to the Accreditation Document", there should be a comprehensive enumeration and exact description of test methods for which accreditation has been granted.

The use of the logo of the accreditation body has to be regulated exactly. When these regulations are contravened, possible sanctions have to be put into motion. Considering the rapidly increasing number of accredited laboratories in Europe, a serious potential for conflict exists caused by vague descriptions of the scope of accreditation and lax intervention possibilities after misuse of the logo; this is true for accredited laboratories as well as for accreditation bodies.

10.10
Surveillance

Surveillance of accreditation has to be given much more weight than may be presumed by the statement of clause 11, EN 45002. Surveillance is decisive for inspiring confidence in constant competent testing. Accreditation of testing laboratories will only gain ground as a confidence inspiring measure if testing competence, assessed at a given time, is constantly maintained and supervised. This, especially, holds true because accreditations effective for up to 5 years are granted as a result of a two days on-site assessment.

Arrangements for surveillance have to be stipulated in the accreditation contract. These provisions authorize the accreditation body at any time, on appointment, to convince itself by an inspection and by suitable control measurements to see if the performance of test methods stated in the accreditation certificate conforms to the standard. The testing laboratory is under obligation to keep a complete record of its calibration measures and to show the records on demand by the accreditation body. The accreditation body must have the authority to inspect test reports and to examine the results of internal quality assurance audits performed by the testing laboratories. As a further surveillance measure for the control of competence, either the accreditation body itself or another body it has designated, has to organize and to evaluate inter-laboratory trials. Proficiency of the testing laboratories has to comply with criteria of the accreditation body declared in advance. When the performance of testing is poor the accreditation has to be withdrawn. The accreditation body must be informed of any important organizational changes and measures related to test activities (rf. sub-clause 7.3 of the guideline). In order to prevent discussions of the significance of a change, it is recommended that the accreditation body should be included in the up-dating service responsible for revision of the quality assurance manual of the accredited testing laboratory.

A basis for trust only becomes established if the commitment of the testing laboratory to keep a constant standard within the scope of granted accreditation is not only tested but if it is also connected to the maintenance and supervision of accreditation.

All WELAC accreditation bodies are unanimous on the surveillance principles. This basic minimum requires further development with regard to special features of testing fields or test items. The consumer protection aspect of accreditation should not be left out of consideration when deciding on the type, scope, and frequency of surveillance measures.

10.11
Accreditation and Standardization

According to the arrangements stipulated by the accreditation contract, the accreditation body controls whether test methods are carried out according to standard specifications.

The accreditation body should also initiate cooperatively the adaptation of standards to actual developments, and, therefore help to ensure the application of standards. Assessing the carrying out of test methods provides additional information for possible improvements in the conduct of test operations according to individual standards. This may supplement the recommendations for making working methods conform to standard specifications as well as the routine examination of standards so that they include the latest technical developments. The accreditation body should pass on related proposals to the relevant standardization organization, and, hence, encourage the modernization of the standards. Testing is only performed according to already accepted standards; this promotes the comparability of test results.

Accreditation, thus, consolidates the acceptance of standards and the position of standardization organizations. Certification of QA systems does not produce such an effect. Work of a certification body is neutral with regard to standards. Thus, discrepancies between testing actions and standards for testing may only be removed by a cooperation between the accreditation body and the standardization organization.

10.12
National and International Agreements on Mutual Recognition

National and European recognition by other accreditation bodies is of great importance for every accreditation body. In Germany, mutual recognition is a prerequisite for membership of the Joint Agencies for Accreditation (Trägergemeinschaft für Akkreditierung, TGA) and the German Accreditation Council (Deutscher Akkreditierungsrat, DAR). At the European level, a mutual recognition of all EC and EFTA accreditation bodies has almost been realized.

Mutual recognition of accreditation systems, however, does not mean compulsion to accept test results of an accredited testing laboratory by third-parties. However, mutual recognition of accreditation systems may facilitate this acceptance. Mutual recognition does not override the market forces in the field of testing; the aim of an accreditation body must be to become firmly established within this mechanism through accreditation competence.

Chapter 11

The Significance of Accreditation in Comparison with GLP

Hendrik Schlesing

11.1
Introduction

More than twenty years ago, the expressed desire for a uniform standard for assessing the quality of investigations louder and louder. At first, this desire was embodied in the statutorily regulated area where, in the beginning, it was essentially restricted to a harmonization of procedures. Only by introducing GLP, was a general measure established which serves, independent of individual procedures, as a valuation basis for the quality of a laboratory. Since GLP has a very restricted application area, consequently, for the accreditation of all testing laboratories, a general measure was developed. In the following, GLP is first introduced, then similarities and also differences between GLP and accreditation will be elucidated and, in conclusion, future prospects and suggestions for further developments will be given.

11.2
GLP – Good Laboratory Practice

11.2.1
Origin

At the beginning of the 1970s, in the United States of America, the FDA (Food and Drug Administration) discovered short-comings when inspecting toxicological trials performed by contracting laboratories. During the so-called Kennedy-hearings authorities, contracting laboratories and pharmaceutical companies were asked for the causes. These hearings led to some proposals for the elimination of those problems encountered. These proposals were adopted by the FDA and summarized in the "Principals for Good Laboratory Practice". The objective of these principals published in 1979 in the Federal Register was, and is, to determine a basis for the conduct of studies in the toxicological sector. Because the publishing of results is compulsory, all companies had to follow these regulations if they wanted to carry out or have studies carried out in the framework of approval, permission, registration, declaration or notification with relation to the Drug Laws of the USA. Hence, since that time state authorities control whether or not a study has been conducted according to GLP.

Because this regulation was applied to all companies importing pharmaceuticals into the US, logically, at international level, the OECD (Organization for Economic Cooperation and Development) concerned itself with the internationalization of this standard. Apart from the main goal of improving test quality, the objective of the OECD harmonization efforts was mutual recognition of tests effected under GLP in the relevant country. After three years of work, in 1981, the "OECD Principals of Good Laboratory Practice" were published and officially recognized by EC member states. In 1987, they were adopted as an EC directive. In parallel to this development, guidelines for GLP study audits were established which, in 1990, were also published as an EC directive. In Germany after a recommendation phase that began in 1983 when the OECD principles were published in the Bundesanzeiger (Federal Gazette), in 1990 the "Gute Laborpraxis" became legally binding through the law on Hazardous Chemicals. In supplementary administration regulations, the guidelines for laboratory inspections were enacted in the same year.

11.2.2
Legal Fundamentals

In article 19a) paragraph 1 of the Law on Hazardous Chemicals, GLP – Good Laboratory Practice – is defined as:

"(1) Non-clinical, experimental studies of substances or mixtures, the results of which are intended to enable an assessment to be made of their hazards to man and the environment by approval, permission, registration, declaration, or notification have to be performed according to the principles of Good Laboratory Practice according to Appendix 1 of this law."

Apart from the law on Hazardous Chemicals, non-clinical experimental studies, as stated above, are prescibed by the following laws:

- Law on Pharmaceuticals
- Law on Plant Protection
- Law on Food and Consumer Goods
- Law on Explosives

Basically, anyone carrying out studies in the above-mentioned statutorily regulated framework, has to comply with GLP principles. The studies themselves are summarized by law in the following categories:

- physico-chemical properties and content determinations
- toxicological properties
- eco-toxicological properties
- behaviour in soil, water and air
- residues.

Anyone capable of furnishing satisfactory evidence that he conducts studies according to GLP principles in a test facility, may apply for a GLP certificate which serves, apart from the legally binding GLP compliance declaration, as a certificate that the study conformed to GLP. The GLP certificate issued by a state authority certifies that the test facility complies with GLP principles in one or more of the above-mentioned categories. The official certificate depends on successfully passing the audit performed by the body designated in the respective Federal State.

11.2.3
GLP Principles

Definition of Good Laboratory Practice:

Good Laboratory Practice is concerned with the organizational process and the conditions under which laboratory studies are planned, performed, monitored, recorded, and reported.

The aim and purpose of GLP is to submit reliable data to authorities assessing these data. These study results constitute the basis for registration of the related test substances; therefore, the results have to be reliable and comprehensive. The studies have to be conducted to a very high standard of quality and the results have to be comparable and traceable, a fact, that demands related documentation. This is the main focus of GLP. By harmonization, data quality is improved and a mutual recognition of data is facilitated; this avoids unnecessary experiments on animals and eliminates barriers to trade.

Section 1 of Good Laboratory Practice defines the terms necessary for its understanding. Section 2 is composed of ten different paragraphs:

1. Test facility organization and personnel
2. Quality assurance programme
3. Facilities
4. Apparatus, material, and reagents
5. Test systems
6. Test and reference substances
7. Standard operating procedures
8. Performance of the study
9. Reporting the results
10. Storage and keeping of records and material.

1. Test facility organization and personnel

Responsibilities within the test facility are determined and guidelines having regard to the following points are given:

- The test facility management is not directly involved in the conduct of studies but has, as its main responsibilities, to ensure that the principles of Good Laboratory Practice are complied with.
- It ensures that when conducting studies all legal regulations are fulfilled,
- ensures that qualified personnel, appropriate facilities, and adequate equipment are available,
- maintains a record of the qualifications of each member of staff (experience, training, as well as job descriptions),

- designates person(s) responsible for quality assurance,
- for each study designates a Study Director and his/her deputy,
- agrees to the study plans and determines, by a standard operating procedure, the guidelines to be followed in the case of amendments to the study plan,
- ensures that standard operating procedures (SOP) are established and followed.

Furthermore, this first paragraph mentions the Study Director's responsibilities including – among others – the overall conduct of the study and its reporting as well as compiling the study plan. Personnel responsibilities, described in the third paragraph, are kept global and are essentially related to using safe working practice when handling substances under investigation or used during the study.

2. Quality assurance programme

Here it is laid down that every test facility should have a documented quality assurance programme. The corresponding responsibilities are determined; essentially, the responsibilities of quality assurance personnel is taken into consideration. Individuals concerned with quality assurance have to acertain that the study plan and standard operating procedures are available to the personnel. They should ensure, by periodic inspections, that study plan and standard operating procedures are followed. Comprehensive records of such inspections should be retained.

3. Facilities

The test facilities should be of suitable size, construction, and location to meet the requirements of the study. More details are given on test system facilities as well as on facilities for handling test and reference substances. Special attention is drawn to suitable archive facilities. Furthermore, this paragraph focusses on waste disposal that is consistent with pertinent regulatory requirements.

4. Apparatus, material, and reagents

Here it is prescibed that apparatus used should be suitable for the study and should be periodically inspected, cleaned, maintained, and calibrated according to relevant SOPs. As in all other paragraphs, here too, keeping records is required. Importance is attached to the fact that materials used in the studies should not impair the test systems. It is important to label reagents appropriately to indicate source, identity, concentration, and stability information as well as to state preparation and date of expiry and – if necessary – specific storage instructions.

5. Test systems

Physical/chemical and biological test systems are specified differently. Considering chemical analysis, the first paragraph provides information identical to the previous paragraph on "apparatus" because here, facts related to obtaining physical and/or chemical data are discussed. The paragraph on biological test systems contains details on environmental conditions, housing, handling and care of animals as well as with importing, collection and care of animals, plants etc. and on the related documentation.

6. Test and reference substances

Records including date of receipt, handling, sampling and storage of the test and reference substances should be maintained regularly. In addition, supplementary characteristic specifications are demanded, i.e. test and reference substances should be appropriatey labelled with regard to:

- Identification
- Batch number
- Composition
- Purity
- Homogeneity
- Stability.

7. Standard operating procedure

The written standard operating procedures should be approved by the management and should be immediately available in each seperate laboratory unit for the activities being performed there. For the following categories of laboratory activities, standard operating procedures should be available:

- test and reference substances
- apparatus and reagents
- record keeping, reporting and storage
- test systems
- quality assurance procedures
- health and safety precautions.

Published method descriptions or the manufacturers' operating instructions may be used, too.

8. Performance of the study

Besides standard operating procedures and reports on test results, a further essential paragraph concerns the performance of the study. Prior to initiation of the study, a study plan should exist in a written form. Study plans should be stored carefully. The study plan should contain at least the following information:

- identification of the test and reference substances
- information concerning the customer and the test facility
- dates

- test methods
- further points (justification for the selection of the test system, character-ization of the test system, method of administration and the reason for its choice, etc.)
- a list of records to be kept.

Considering the conduct of a study, it is regulated that to each study a unique identification should be given and that all items concerning the study should carry this identification. Furthermore, a comprehensive documentation is required to demonstrate that the study has been conducted in accordance with the study plan, that all related data have been recorded, and that all entries have been dated, signed or initialled.

9. Reporting the results

For each study a final report should be prepared. In this paragraph informa-tion is given on the content of the final report:

- identification of the study, the test and reference substances
- information concerning the test facility
- dates on which the study was initiated and completed
- quality assurance statement
- description of materials and test methods
- results
- storage.

10. Storage and retention of records and material

The last paragraph describes the way archives should be equipped and what items have to be stored:

- study plans
- raw data
- final reports
- report of laboratory inspections and study audits
- samples and specimens.

Furthermore, only authorized personnel should have access. The second paragraph states that, beside the above mentioned documents, also the fol-lowing have to be retained: a summary of qualifications, training, practical experience and job descriptions of the personnel as well as the records, and reports on the maintenance and calibration of equipment, and, further-more, a record of standard operating procedures for 30 years after signing the final report. An exception is made for samples and specimens that should be retained only as long as the quality permits evaluation; but, at least for 12 years.

   The archive, like the quality assurance unit, but independent of this unit, is under direct control of the test facility management.

11.2.4
GLP Certificate

Laboratories, conducting studies in accordance with GLP, are submitted to compliance monitoring by authorities for the purpose of verifying adherance to GLP principles. As already indicated by the Law on Hazardous Chemicals, on 29.10.1990 the corresponding administrative prescription GLP came into force, stipulating guidelines for the conduct of test facility inspections and study audits. The laboratory inspection procedure starts with a pre-inspection. This pre-inspection lasts no more than one day and serves to familiarize the inspector with the test facility and to discuss those documents he had ordered for preparatory work. If, already on application, documents were furnished, these will be discussed now. Date and time of the visit are scheduled, scope and personnel involved are identified. The laboratory inspection lasts about one day or more, depending on the test facility. An extensive examination has to verify if the test facility's activities adhere to the principles of GLP. The laboratory inspection consists of several phases: starting conference, examination of documents, study audits, and the laboratory inspection itself and the inspection of the archives. Subsequently, in a final discussion deviations from GLP principles found in the course of the visit are summarized. Minor deviations are orally communicated, serious deviations identified are reported in a short written protocol signed by all individuals involved. Under some circumstances, depending on the report on non-compliance, there may be a need to revisit the laboratory to verify that action has been taken to remedy deficiencies. The inspection report contains, for each item, a detailed description of the test results stating all discussion points as well as the final vote of the inspection team with regard to the issue of the GLP certificate.

11.2.5
Personnel

As stated several times in the principles of Good Laboratory Practice, personnel and their qualifications and training play an important role. Appropriate qualifications have to be considered. However, this alone is not sufficient. Additionally, a minimum of staff resources is necessary for the required implementation of internal quality assurance. Apart from the test facility manager and deputy, a person responsible for the archives and deputy, a person responsible for quality assurance and deputy, a Study Director with deputy, and a technical employee and deputy are needed. This indicates that small facilities may well have problems covering all this positions.

11.2.6
Time Needed

The Implementation of Good Laboratory Practice in a medium-sized institute with a staff of about 30 to 50 people will demand about one man-year of work.

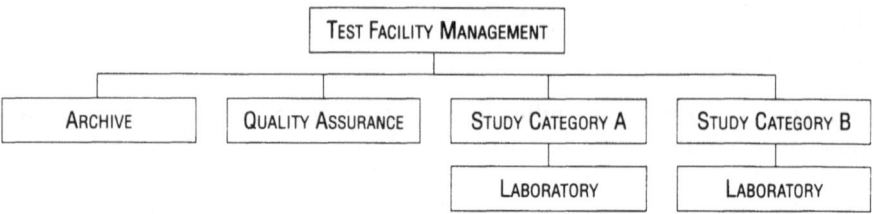

Fig. 1. Organizational chart of "GLP"

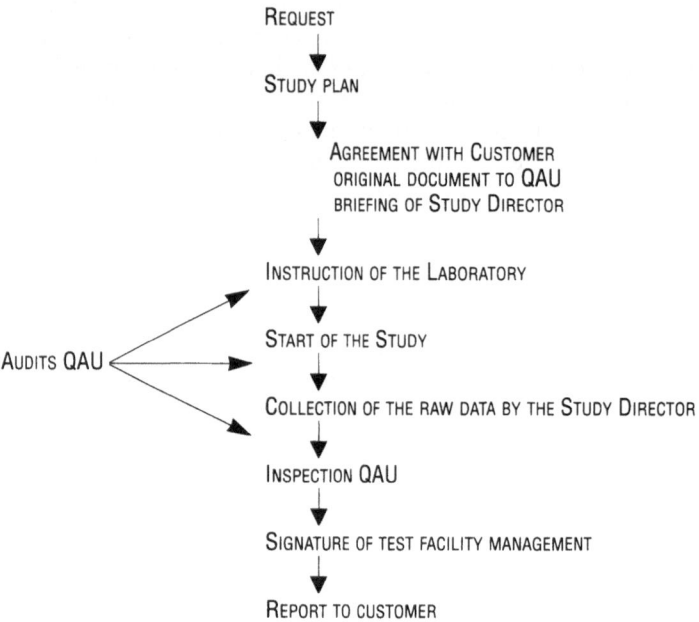

Fig. 2. Course of a GLP study

A number of quality assurance criteria, required in the framework of GLP and already present in modern laboratories, help to keep within certain limits the additional expenditure that arises in comparison with studies not conducted under GLP. Hence, the additional expenditure is certainly acceptable.

11.3
Accreditation

Previous chapters of this book have reported extensively on accreditation. Therefore, prior to comparing GLP and accreditation, only some advantages and disadvantages of accreditation are presented here.

The main advantage of accreditation is the fact that it enables one to compare the total quality level of laboratories. In the regulated area, various criteria or assessments of laboratory quality are known. A uniform international measure was lacking. GLP was the first approach; however a uniform regulation is only now given through accreditation. A comparison of recognition practiced in the regulated area of individual states of the Federal Republic reveals very clearly that there are enormous differences; this, even in the case of audits that should be performed uniformly (e.g. the designation of a testing body according to the regulation on Hazardous Substances or Federal Emmission Law, Drinking Water Act, etc.).

Another advantage is that accreditation makes entry to the European internal market as well as export to third countries much easier. Importing and exporting companies now can rely on uniform principles and thus repetitive testing is avoided.

The first and most obvious disadvantage with regard to accreditation is the costs arising for accreditation itself as well as those arising for continuous surveillance and, later on, for the participation in interlaboratory studies. This is only a feeble argument because every laboratory should meet certain minimum measures with regard to quality. A considerably more persuasive argument that an additional bureaucracy has been built up with concurrent accreditation bodies that have to be supported.

11.4
Comparison of GLP and Accreditation

Both have the supreme goal of achieving a comparable quality of laboratory results on an international level. This is attempted by implementing uniform assessment criteria, stipulating regulations for the type and content of the laboratory documentation system, the laboratory organization, the structure of the laboratory quality assurance unit and its tasks.

First, the main difference arises from the fact that GLP is legally binding (Law on Hazardous Chemicals). This means that everyone conducting studies in his test facility within the legal framework has to adhere to GLP principles. The test report includes a certificate, a legally binding GLP compliance declaration for each study as well as the GLP compliance certificate issued by federal authorities. In addition, everybody who can prove lawful interest may claim such a certificate.

In contrast, a testing laboratory wishing to initiate an accreditation according to EN 45000 may not enter a claim here in the statutorily non-regulated area. Accreditation is the result of a private agreement between the accreditation body and the laboratory. There is no evaluation of the needs. On the other hand, a common feature for both is control by the authorities (GLP) every two years or for a period fixed by the accreditation body.

A survey of differences and common features is given Table 1. That one is regulated by law and the other not is the most obvious difference between them.

Table 1. Differences/similarities of GLP/EN 45001

	GLP	EN 45001
Scope	Lawon Hazardous Chemicals	Not regulated statutorily
Objective	Comprehensibility through complete documentation	Comparability of analytical results
Organisation of the laboratory	Completely defined Study plan for each study	Recorded test methods
Quality assurance	QA programme with independent QAU	QM system with QA Manual for *all* personnel
	Only internal QA	Also external QA (round robin)
Laboratory/test facility management	Separation of tasks QA/test facility management	Management also supervises QA system

Apart from this, many common features exist which are expressed by their common goals, i.e.

– the improvement of laboratory quality, and,
– the improvement of the transparency of analytical results.

For example there are similar requirements for personnel qualifications and training.

Requirements concerning documentation are alco comparable – with the restriction that storage and retrieval with its extremely long period for retaining data requires essentially higher expenditure in the case of GLP. The standard EN 45001 only stipulates that the period for keeping the documents should adapted to the requirements and to regulate this fact in the quality assurance manual.

Requirements for premises and equipment are treated in a similar way, while GLP imposes detaile requirements because it covers toxicological studies (biological test systems). Both standards attach similar requirements to the handling of test items.

In many cases, differences between GLP and the standard are only given by details of the requirements. The standard, for example, focusses on the use of reference materials and on their traceability to national or international standardized materials. GLP does not make these requirements, but all information is gathered in order to identify and characterize the reference substances.

Both require the documentation of records (standard) or raw data (GLP) which should guarantee traceability or comprehensibility of test data obtained. In the framework of GLP, a slightly more stringent regulation for the determination of raw data is applied.

In principle, sub-contracting is possible with regard to both regulations. According to GLP, the sub-contractor has to be GLP-certified. According to the standard, the sub-contracting laboratory has to comply with the requirements, but there is no obligation for accreditation.

The content of the reports destined for the client is comprehensively described in both listed regulations. In contrast to GLP, the standard states that the test report for the client must not include valuations or judgements. For GLP, on the other hand, in particular, the Study Director's compliance declaration and the quality assurance declaration of the quality assurance unit are absolutely required. Additionally, the test report should incorporate all further information which may have an influence on the quality and integrity of the study. Such stringent requirements do not exist for the standard.

There are two aspects where differences between standard and GLP become most intense:

– quality assurance, and,
– study plan.

11.4.1
Quality Assurance

In EN 45001 no similar requirement is found for the GLP requirement for a quality assurance programme that details the tasks and function of an independant quality assurance unit (QAU) which should monitor the compliance to GLP principals by periodic inspections.

Because of its wide-spread application area, the standard requires a quality management system that is adapted to the relevant testing. The elements of the quality management system are described in the quality manual. This manual is intended for use by all personnel and not, like the GLP quality assurance programme as an instruction for the exclusive use of quality assurance personnel. The quality manual is written in detail and contains all essential characteristics for an assessment of systematic assurance and an auditing of the reliability of individual procedures. Another important difference is, as already mentioned, that there is no need for special quality assurance personnel; a fact which certainly has a positive effect on analysis costs.

11.4.2
Study Plan

A special feature of GLP is the study plan which defines the entire scope of each study. Similar requirements are not claimed by the standard. A link may be presented by the standard operating procedures required in the frame of GLP, which may be compared with the documented test methods required within the scope of the standard. The client has to agree to the study plan in writing. He thus confirms the planned extent of the study. Here, one may compare to the standard clause defining conditions for the coopeartion with clients. This is confirmed by the requirement that the client's request has to be forwarded in a clear and precise way to the laboratory and that a defined complaint procedure is required in order to clarify, if necessary, eventual disagreement with the client. While the standard affords the client access for observing a test performed for him, this is not in GLP regulations, but has, in fact, for many years been handled by the clients by performing audits regularily.

11.5
Summary and Future Trends

First, it has to be clearly stated that, as a common feature, GLP and the European Norm have as their primary objective the realization of comparability and reliability of test and analysis results. In addition, there has been a general increase of the quality level which, in many cases, was certainly necessary. Last not least, for the elimination of trade barriers: mutual recognition of the corresponding test and analytical results. The starting point of GLP was a need for regulation in the statutorily regulated domain. Statutory regulations had to be enacted, because here, the type and extent of studies are very precisely defined and authorities are involved in assessing the risks chemicals carry for man and the environment. Compared to the standard, GLP thus defines, in detail, regulations specific for toxicological studies. GLP is not yet in its final stages because its extension to e.g. field studies is pending. This has already been documented by related OECD publications and even been realized in some cases.

The standard covers a far larger area of testing and therefore its criteria have to be considered as being general criteria. The starting point was to define a guideline which should represent a quality management system that relies strongly on the laboratory's own responsibility. The quality management system required in the framework of the standard has to be adapted to the testing laboratory's special needs and tasks and to the required testing.

It is important to mention, that both regulations, first considered as competing, should not be considered as such, but should possibly complement each other.

Certainly, regarded as the main standard, ISO 9000 may serve as a basis for quality improvement or quality assessment of testing laboratories. Special forms like EN 45001 and GLP may be considered as being derived from ISO 9000. It is desirable to collate closely the different standards which then are really complementing one another. They may then be used, like cascades, to respond to the testing laboratories requirements that become more and more specific.

Taking into account approval/registration, necessary in the statutorily regulated area for many years, and the related requirements, one has to strive towards, e.g. starting with ISO 9000, that all recognitions agreed upon and various certificates will enjoy mutual recognition. In the statutory domain this will eliminate re-testing actually performed by the authorities of the States. Here, obviously, less expenditure would result for the laboratories concerned. Corresponding discussions have been initiated between various committees.

Globally stated, GLP has certainly exerted and still has a positive influence on the definition of EN 45001. The same is true with regard to the conduct of audits. An improvement of subjective and objective quality as well as of quality consciousness of laboratories has already been realized. In this way, one of the main purposes of the two regulations has already been reached.

EURACHEM – Organization for the Promotion of Quality Assurance in Analytical Chemistry and the Accreditation of Analytical Laboratories in Europe

Helmut Günzler

12.1
The Founding of EURACHEM

As mentioned in chapter 1 (H. Berghaus) of this book, in 1985, the Council of the European Community recognized that differences in national accreditation systems was a hindrance to trade for the internal market. The Council decided at that time on a new concept for harmonization and standardization whose aim was to facilitate the mutual recognition of testing and certificates. The "Global Concept" which developed from this, embracing both the statutorily regulated area and the free market economy, is concerned with the promotion of mutual trust through transport competence and quality.

With the establishment of *Quality Management Systems (QMS)*, the reliability of analytical results is supported in the same way as we have met in the manufacture of products for some years. A QMS summarizes all the structures, responsibilities, procedures, processes, and methods needed for achieving the quality of a product or service (e. g. chemical analyses) agreed on or necessary for a specific purpose (cf. chapter 3). These measures are set down individually in a *Quality Manual (QM)*. A QMS as a system and a QM as a document can be certified according to ISO 9000 or EN 29000. In analytical chemistry, the proof of efficient quality assurance is the result of the particular laboratory undergoing a process of *accreditation,* i. e. the *proof of competence for the carrying out of specific analytical methods in comparison with an independent third party.*

The general criteria for the operation of testing-laboratories, for their inspection as well as for the organization of accreditation bodies are regulated by international standards [1, 2]. These have been interpreted for analytical laboratories [3].

The aim of accreditation policy is to promote the reliability of testing and analytical results within their sphere of activity – which in the long run should in no way remain confined to the European market – to create mutual trust in results and so render expensive retesting (by manufacturers of products and the receivers of the goods) superfluous. The question remains open, however, as to whether the companies concerned (receivers of goods and their internal testing laboratories), the authorities, and other institutions will gain trust in the efficient working of the system just through the existence of the standard.

In order to find a solution to this problem, in 1988, analysts from several European countries, came together. Soon delegates from analytical laboratories, bureaus, authorities, universities and of industry of all EC and EFTA countries joined the meetings of this group and estabished a committee under the name of EURACHEM (Co-operation for Analytical Chemsitry in Europe) which was constituted by signing a memorandum of understanding in 1990.

12.2
Objectives of EURACHEM

EURACHEM was created to enhance confidence in the results produced by analytical laboratories. It provides a network for collaboration between European laboratories and other key organisations. EURACHEM acts as a forum for the discussion of common problems and the formulation of considered views on policy, organizational and technical issues. It provides a framework within which European laboratories can decide priorities and compare results.

EURACHEM works with existing national and international organizations and builds on the strong sectoral and informal links which exist between analytical chemists. Avoiding overlap with existing initiatives and areas of responsibility will be a priority. EURACHEM has the aim of promoting:

- awareness of analytical quality problems;
- analytical quality management based on EN 45 000, ISO Guide 25 and Good Laboratory Practice standards;
- measurement traceability underpinned by reference materials;
- performance testing including the establishment of internationally agreed guidelines for operating schemes.

Business meetings are supplemented by technical seminars and workshops. Communication with the wider analytical community is facilitated by the publication of news in the chemical literature.

12.3
Structural Organization of EURACHEM

The EURACHEM committee consists of two delegates from each EU and EFTA member state as well as from the Joint Research Centre. States not within the EU may acquire the status of associated member on request.

In most of the member states, national bodies have already been constituted or are going to be. To promote information transfer between the international body and the laboratories, they also distribute documents prepared within EURACHEM (guidelines, documents) such as interpretations of standards, guidelines and quality assurance elements.

12.4
Tasks

Starting with these objectives, EURACHEM defined its tasks which are undertaken in different working groups:

– interpretation and further development of the EN 45000 series of standards: The standard EN 45001 only providing *general* criteria for the operation of (all types) of testing laboratories, a working group founded together with EAL (formerly WELAC) prepared an interpretation with respect to chemical laboratories [1]. A further EURACHEM/EAL-working group and a EURACHEM-working group are actually working on interpretations of EN 45001 with regard to microbiological laboratories and with regard to laboratories performing non-routine analysis as well as analytical development work, respectively. Furthermore, based on the experience gained up to now in field laboratories, EURACHEM intends to influence the further development of the EN 45000 series.
– EURACHEM activities are mainly devoted to the discussion of all questions and problems related to quality assurance in analytical chemistry and the consolidation of knowledge regarding its fundamentals. In this context, EURACHEM organizes international workshops which serve to determine the current know-how, to identify controversial questions and as yet unsolved problems, and to work out solutions or procedures. The results of these workshops are summarized and documented by working groups in order to be distributed by national committees to the practising analyst of laboratories throughout Europe.
  The workshops centre on:

– quality assurance in the analytical laboratory;
– comparability and traceability in chemical measurements;
– measurement uncertainty;
– proficiency testing.

However, validation, reference materials, metrology and calibration are also elements of quality assurance the substantial content and practical application of which lies in the hands of experts in the working groups.

– A particularly important aspect among the activities of EURACHEM is to introduce knowledge of quality assurance in analytical chemistry into tuition and training. In a few years, no analyst should leave university without sufficient basic knowledge of quality assurance. The organization of national training courses and seminars is a prime task of the national EURACHEM committees in order to promote quality awareness within analytical chemistry.

In accordance to these responsibilities, EURACHEM is sub-divided into a series of working groups which exist either permanently or temporarily until the accomplishment of their task. Once a year, all delegates gather for the

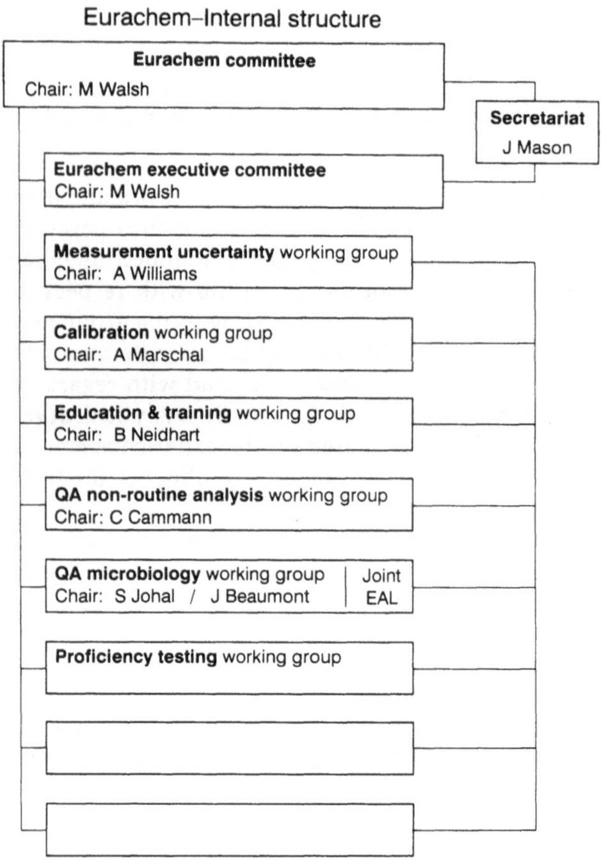

Eurachem–Internal structure

```
┌─────────────────────────────────────────────┐
│             Eurachem committee                │
│  Chair: M Walsh                               │──────┐
└─────────────────────────────────────────────┘      │ ┌──────────────┐
    │                                                 │ │ Secretariat  │
    │ ┌─────────────────────────────────────────┐     │ │ J Mason      │
    │ │      Eurachem executive committee        │─────┘ └──────────────┘
    │ │  Chair: M Walsh                          │
    │ └─────────────────────────────────────────┘
    │ ┌─────────────────────────────────────────┐
    │ │ Measurement uncertainty working group    │
    │ │  Chair:  A Williams                      │
    │ └─────────────────────────────────────────┘
    │ ┌─────────────────────────────────────────┐
    │ │ Calibration working group                │
    │ │  Chair:  A Marschal                      │
    │ └─────────────────────────────────────────┘
    │ ┌─────────────────────────────────────────┐
    │ │ Education & training working group        │
    │ │  Chair:  B Neidhart                      │
    │ └─────────────────────────────────────────┘
    │ ┌─────────────────────────────────────────┐
    │ │ QA non-routine analysis working group     │
    │ │  Chair: C Cammann                        │
    │ └─────────────────────────────────────────┘
    │ ┌──────────────────────────────┬──────────┐
    │ │ QA microbiology working group │ Joint    │
    │ │ Chair:  S Johal / J Beaumont  │ EAL      │
    │ └──────────────────────────────┴──────────┘
    │ ┌─────────────────────────────────────────┐
    │ │ Proficiency testing working group         │
    │ └─────────────────────────────────────────┘
    │ ┌─────────────────────────────────────────┐
    │ │                                           │
    │ └─────────────────────────────────────────┘
    │ ┌─────────────────────────────────────────┐
    └─│                                           │
      └─────────────────────────────────────────┘
```

Fig. 1. EURACHEM – Internal Structure (Nov. 1995)

Committee Meeting where all important decisions are taken. An Executive Committee organizes these meetings and controls the activities of the committee (Fig. 1).

12.5
Cooperation with Other Committees

Striving for quality assurance in all areas of commerce and industry has brought up a diversity of new committees which are directly or indirectly involved with accreditation and certification. Even the expert in this field has difficulties in identifying all the organizations which are hidded behind more or less exotic acronymes. The "educated nonexpert", however, loses track of things. Therefore, the appendix to this volume provides a list of organizations which are actually operating in Europe in the fields of accreditation, certification, and measurement and testing; their main tasks and the year of foundation are indicated as well. Most of these organizations were founded during 1987 and 1993. Only the cooperation of calibration services (WECC) and the accreditors of laboratories (ILAC) have maintained, since the mid 1970s, international cooperation and a permanent international exchange of experience.

EURACHEM maintains close contacts with these organizations, in general by delegating members of the Executive Committee, with the intention of either avoiding all duplicate work or, at least, working cooperatively on similar projects. So there is particularly intense cooperation with EUROLAB, that has been institutionalized by memorandum of understanding; EUROLAB pursues the same objectives for all remaining sectors of commerce and industry as EURACHEM does in the field of chemistry. In 1994, a welcome example for a reduction in the multitude of organizations was the fusion of WELAC and WECC into EAL (see appendix).

The transnational significance of the comparability of analytical results – a result of successful quality assurance – has been unequivocally recognized in the context of the implementation of the EU Common Market; however, also indisputably, this is not just a continental concern but also a global one. Therefore, cooperation of the European committees with corresponding organizations in the American and Asian economic domains is absolutely necessary (refer to chapter 7). EURACHEM therefore emphatically supports worldwide cooperation, as for example ILAC, and has actively used its influence to further the foundation of CITAC.

12.6
Summary

EURACHEM serves as a forum for the discussion of all questions arising within the context of accreditation, certification, quality assurance and chemical metrology. The committee was constituted in 1990 in order to promote the harmonization of national accreditation systems in Europe and the inspiration of mutual confidence in analytical results by promoting contact between chemists/analysts of industry, authorities and university.

EURACHEM consists of two delegates from each EC and EFTA member state; on request, states of Central and Eastern Europe may obtain the status of associated member.

EURACHEM promotes knowledge in all provinces of quality assurance in analytical chemistry and its introduction into the curricula of universities by organizing international workshops and establishing working groups entrusted to perform special tasks. It is actively involved in the interpretation and further development of the international series of standards for accreditation and certification.

EURACHEM maintains close contacts with all relevant committees and organizations and strives for worldwide international cooperation in the area of accreditation, certification, and quality assurance in analytical chemistry.

12.7
References

1. CEN/CENELEC (1989) European Standard EN 45001: General Criteria for the operation of testing laboratories: Brussels, Belgium
   CEN/CENELEC (1989) European Standard EN 45002: General criteria for the assessment of testing laboratories; Brussels, Belgium
   CEN/CENELEC (1989) European Standard EN 45003: General Criteria for laboratory accreditation bodies; Brussels, Belgium
2. International Organization for Standardization (1990) ISO/IEC Guide 25: General requirements for the technical competence of testing laboratories; Third Edition, Geneva
3. WELAC/EURACHEM Accreditation for Chemical Laboratories: Guidance on the Interpretation of the EN 45000 Series of Standards and ISO Guide 25

Chapter 13

The Accreditation of Environmental Laboratories in the United States

E. Ramona Trovato, Jeanne Mourrain, Joseph L. Slayton*

13.1
Introduction

13.1.1
Monitoring Systems

The United States environmental protection Agency (EPA) relies heavily upon analytical data in decision making processes to carry out its mission of protecting environmental resources and safeguarding public health. These data result from a great many analyses and, it has been estimated that the environmental testing business grossed over $ 2 billion in the U.S. in 1993 [1, 2].

The analytical data employed by the Agency may be divided into two broad categories: compliance monitoring and ambient monitoring. Compliance monitoring refers to required sampling and analyses under regulations governing pollution sources. Ambient monitoring includes environmental sampling and analyses to establish baseline or reference background levels and trends. Generally, these data are used to ascertain environmental quality.

Compliance monitoring includes analyses performed directly in EPA and state laboratories, as well as by the regulated community. In fact the majority of the data are based on self-monitoring [3]. Under these programs, permits are issued to facilities (industries, waste treatment plants, drinking water production facilities, hazardous-waste sites, air pollution emission sources, etc.) and limits are set with regard to various pollutants. The required frequency and method of analysis, as well as, allowable pollutant discharge limitations are generally delineated in a permit. The permit serves as a legally binding agreement between the regulated community and the regulator. Sampling and analyses specified by the permit are performed by the permittees. The mandatory sampling and analysis procedures are specified in detail and published in the Code of Federal Regulations (CFR). This monitoring is conducted by permitted facility personnel or may be performed by commercial consulting firms

---

* Disclaimer:
  The mention of trade names, organizations or products is for illustrative purposes and should not be considered on endorsement by the U.S. Environmental Protection Agency. This material reflects the opinions of the authors and has not been reviewed or endorsed by the EPA.

and laboratories employed by the permittee. Oversight of the quality and effectiveness of the compliance monitoring activities is the responsibility of the EPA and each state. Compliance data are routinely reviewed for conformity with the issued permits by program specific oversight units. Also, on-site inspections are routinely performed to assure compliance with self-monitoring sampling, analysis, and reporting requirements of the permits.

In addition to compliance monitoring programs, a great number of analyses are performed to ascertain the health and quality of the environment through ambient monitoring programs. Programs to monitor general water quality (ground and surface) and air quality are conducted by most states, often with supporting funds provided by EPA grants. Also, numerous monitoring programs may focus on a particularly important environmental resource, e.g., an estuary such as the Chesapeake Bay or a geographic area of significantly elevated environmental or human halth risk. Such focused ambient monitoring projects are often large and complex, involving thousands of samples and analyses and millions of dollars in annual support. Oversight of the quality and effectiveness of ambient monitoring is the shared responsibility of each state and the EPA. Large, on-going and focused monitoring projects are often overseen by dedicated state and/or EPA personnel. Unlike the highly regulated and restrictive mandates governing sampling and analysis procedures under compliance monitoring, ambient monitoring is generally governed by general guidance. This guidance is provided by the program providing the funding for the project and involves the review of study proposals, which address the scope and objectives of the study.

Many compliance and ambient monitoring programs rely on the analysis of "Performance Evaluation (PE)" samples (often referred to as proficiency test samples) as a tool to help determine laboratory performance and to estimate the data quality. PE samples generally have a known "true" or theoretical value, which is well supported and documented by repeated study and analysis. These are distributed to laboratories as unknown concentrates in sealed glass vials. Though not "double blind", such samples have proven challenging to environmental testing laboratories and are a valuable tool for use by the regulating programs to estimte laboratory performance and data quality, e.g., estimated precision and bias associated with the measurement of a given pollutant. The general assumption is that laboratory performance under such test conditions reflects the best performance possible by the laboratory.

13.1.2
Challenges

These diverse compliance and ambient monitoring programs are controlled by equally complex state and EPA oversight systems. Each of EPA's programs, (e.g., water, air, solid waste, ambient monitoring), has its own unique organizational unit to oversee analytical quality, as to the corresponding programs in each of the state organizations. In addition, the EPA and State have significant differences in the level of detail required in the review of a laboratory. This has resulted in concern for uniformity of quality among the various sources of environmental data. Some laboratories have never been inspected and yet other laboratories have been inspected many times by each program. Laboratories may be inspected for: wastewater analyses (under the National Pollutant Discharge Elimination System of the Clean Water Act); drinking water analyses (laboratory certification under the safe Drinking Water Act); solid and hazardous waste analyses (under the Resource Conservation and Recovery Act (RCRA) and the Comprehensive Environmental Response, Compensation, and Liability Act (CERCLA); air analyses (under the Clean Air Act); radon analyses (under the Superfund Amendments and Reauthorization Act); and toxic substances analyses, including pesticides, fungicides, asbestos, and lead (under the Toxic Substances Control Act (TSCA) and the Federal Insecticide, Fungicide, and Rodenticide Act (FIFRA)).

Mutual acceptance of laboratory inspections, reciprocity, among states of EPA Agencies has been minimized by the diversity of requirements. Specific requirements under the CERCLA' Contract Laboratory Program (often termed Superfund) are markedly different from the RCRA requirements, although both programs are within the same EPA office. These requirements are in turn markedly different from those delineated as Good Laboratory Practices under TSCA and FIFRA, which stress the importance of complete documentation. These requirements are different from the laboratory certification requirements under SDWA, which are also different from the procedures mandated for compliance monitoring for wastewater and sewer discharges for NPDES under the Clean Water Act.

The complexity and diversity within the EPA is mirrored in the state agencies [4] and amplified by the fact that there are 50 states. Each of the state programs has its own emphasis and characteristics and represents an interpretation of the stature and EPA guidace. In addition, each state has the option to be more restrictive or demanding than required by the corresponding Agency guidance or the law. Diverging requirements among various states often make reciprocity difficult or impossible. A commercial laboratory wishing to perform analyses in a given program area (e.g., drinking water) must be inspected by each state in which it has clients unless such states have reciprocal agreements to accept each others assessment of laboratory and data quality. Under this system confusion may reign. Even the loss of a laboratory's accreditation status for one state may be unknown to other states for which the laboratory currently holds certification.

In addition to this already complex and tangled monitoring oversight system, numerous federal, state, and local government programs provide accreditation. Accreditation has been defined as the criteria for evaluting the operation of laboratories and denotes formal recognition that a laboratory is competent to carry out specific tests or specific types of tests [5]. It has been estimated that over 70 state accreditation programs evaluate drinking water, non-potable water, air emissions, and solid and hazardous waste [6]. Examples of such programs include: the National Institute of Science and Technology (NIST), "National Voluntary Laboratory Accreditation Program" (NVLAP)

[7]; the New York Health Department, "Environmental Laboratory Approval Program"; and the New Jersey Department of Environmental Protection, "Water, Testing Laboratories Accreditation, Quality Assurance Program".

As with the other aspects of the oversight systems for data quality, PEs are issued separately under each program. This results in a scheduling nightmare for the environmental laboratories and yet the analytes and analytical protocols are often redundant resulting in wasted time and money. Most PE programs involve multiple testing during the year, e.g., Safe Drinking Water Act PEs involve two concentration levels per study and two studies per year. A "not acceptable" result for either concentration is considered unsatisfactory and participation in the next PE study within the year is mandatory. Laboratories are repeatedly tested during a given year for the same analytes by a number of programs. Yet other analytes are not evaluated with PE's at all, e.g., semivolatile organics are not included in the PE survey for wastewater.

13.1.3
Concerns for Data Quality

Adding to the concerns about data quality is the reality that most state organizations do not accredit for all environmental programs and are frequently limited to drinking water certification (New York and California are notable exceptions). Not only does this drive commercial laboratories to seek accreditation with other states but also from private (non-governmental) organizations. In both cases this accreditation is expensive often requiring the facility to pay large fees, purchase PE samples and pay for the travel expenses of the onsite inspectors. Examples of private accrediting organizations include; the American Association for Laboratory Accreditation (A2La) [8], the American National Standards Institute (ANSI) [9]; the American Industrial Hygiene Association (AIHA); the American Society of Quality Control (ASQC); and the National Sanitation Foundation (NSF).

Given that EPA, state, and commercial organizations, each have different requirements, a commercial laboratory in the United States, spends considerable time and effort participating in numerous on-site reviews and PE studies. Many such laboratories have staff members who specialize in interacting with various inspectors: provide the necessary information prior to the inspection; coordinate any special requests, such as performing a specific analysis on a given day during the inspection; accompany the inspectors during an on-site inspection; attend an inspection "exit debriefing", and provide any required follow-up communication, which generally involves responding to a list of required and suggested changes in the laboratory operations.

Despite the problems of redudancy of inspections, and the nonuniformity of the inspection requirements, many laboratories are only inspected for a small proportion of the services they provide and other labs are not inspected at all [10]. Laboratories not inspected often have a competitive edge because they avoid costly quality control and other procedures which were required of audited laboratories. This may result in the laboratory with the minimum of substandard quality control programs, equipment or analytical procedures under-bidding and adversely affecting the quality of the data supporting a given monitoring program. Also, the results of laboratory reviews from on-site program inspections are not readily available to other clients of the laboratory. Similarly, if one organization has found analytical difficulties associated with a given laboratory, this information is not readily available to other organizations, even if the analyses are similar or identical.

The problems for the United States environmental laboratory community are further compounded by competitive pressure to be accredited by international organizations, e.g., the International Organization for Standardization (ISO [11, 12]. For commercial laboratories in the United States to be recognized in foreign analytical markets such requirements as delineated in the ISO Guide 25 "General Requirements for the Competence of Calibration and Testing Laboratories" must be fulfilled.

13.2
Policy Development

13.2.1
Background

The need for quality data and the disorganized approach to laboratory accreditation led EPA Deputy Administrator F. Henry Habicht II to charter the Environmental Monitoring Management Council (EMMC) in February of 1990 to develop, evaluate, and recommend policy on environmental monitoring activities. Among the issues examined by the EMMC was the feasibility and advisability of establishing a national environmental laboratory accreditation program. EMMC consists of senior management officials representing all EPA programs (e.g., drinking water, water pollution control, solid waste, hazardous waste, air, radiation, pesticides, and toxics). Under EMMC's direction, an Ad Hoc Panel was established to investigate the feasibility and advisability of a national environmental laboratory accreditation program.

The first report of the Ad Hoc Panel identified numerous benefits of a national program, including benefits to the states, the laboratory community, and the regulated community. In particular, a national program that achieves reciprocity among state certification programs would eliminate the duplication that currently exists among public and private programs. The EMMC concluded that three were benefits to a national program and that broad participation by all affected interests was necessary in order to best characterize the needs of the user community, recommend a preliminary program design, and achieve broad consensus. Therefore, in July 1992, the Committee on National Accreditation of Environmental Laboratories (CNAEL) was chartered. CNAEL was composed of representatives from the laboratory and regulated industry communities, other federal agencies, the states, public environmental interest groups, academia, and private accrediting bodies.

EPA's charge to CNAEL was: 1) to determine if there was a need for an environmental laboratory accreditation program and what advantages would be derived from establishing a program; 2) to identify options for operating a national program; 3) to identify other alternatives to a national program which would address the needs of the affected groups; and 4) to recommend an appropriate role for EPA in developing or implementing any program. Each group represented on CNAEL presented their persepctive which included descriptions of the various accreditation (or evaluation) programs currently operated by the states, federal agencies, industry, and private accrediting bodies.

13.2.2
Initial Perspectives

Many federal agencies are both generators and users of environmental monitoring data. For example, the Department of Defense is prepared to operate without a national program but believes that the work could be accomplished more efficiently and at lower cost to the government if a national program were established.

Industry favors the establishment of a national program because of the lack of readily available information on current accreditation programs, the increased costs which are passed on to the consumers, and the disappointments industry has experienced with some of the existing programs.

The laboratory community also supports national accreditation for the benefits which might accrue to the laboratory under such a system: fewer but more comprehensive audits, fewer redundant performance evaluation samples, supgrading of quality assurance programs in all laboratories providing a level playing field in the economic arena, and nationwide recognition of performance capabilities.

The private laboratory accreditors prefer a national program which is consistent among all states and private parties because there are major differences in the thoroughness of existing programs. As a result private industry or other laboratory users are forced to conduct their own evaluations. The limited scope of existing programs further exacerbates this problem. A uniform national program should also ensure that U. S. laboratories are accepted throughout the world.

States originally began establishing independent accreditation programs because there was no other surce. One reason for the inconsistencies among state programs is, in part, that no guidance existed at the time most of the programs were established. The cooperation of the states is needed to assure a successful national program.

13.2.3
Assessment of the Need for a National Environmental Laboratory Accreditation Program

CNAEL conducted a needs assessment based upon a review of the existing literature, work already conducted by EPA and other groups, and an analysis of comments and information provided by CNAEL members and the public. CNAEL analyzed and prioritized the needs, then converted them into a list of goals. The goals were as follows:

- Facilitate reciprocity;
- Standardize sampling, analytical methods, and quality control;
- Provide an objective evaluation of laboratory performance in various sample matrices, including provision of proficiency test samples;
- Establish a comprehensive and flexible program;
- Eliminante redundancy and inconsistency in the current system;
- Promote communication between user and provider;
- Identify and prosecute fraud quickly and justly;
- Allow for improved/new techniques in a timely manner;
- Establish a program which is practical in all aspects;
- Establish uniform inspections/audits;
- Produce data of known, legally defensible quality;
- Require uniform standards for analysts;
- Assist data users to select laboratories;
- Assist laboratories in identifying and correcting problems;
- Provide timely technology transfer and methods and quality control requirements.

These were distilled into a single goal statement: *To obtain data of needed quality in a cost effective manner.*

13.2.4
Evaluation of Alternatives to National Environmental Laboratory Accreditation

CNAEL identified and evaluated potential alternatives to national environmental laboratory accreditation. The 15 alternatives identified by CNAEL in the order of preference were:

- National Environmental Laboratory Accreditation
- Performance Evaluation Testing
- Federal Pre-emption
- Laboratory Process Certification
- Quality Assurance Systems Certification
- National Guidelines
- Reciprocity
- Resident Inspectors
- Training
- Analyst Certification
- Product Certification
- Fraud Audits
- Laboratory Manager Certification
- Performance Bonds
- Status Quo

The following subset of the alternatives was agreed upon to be carried forward in the analysis:

- National Environmental Laboratory Accreditation (includes on-site audit)
- Performance Testing + QA Systems Certification + Lab Process Certification (includes on-site audit)
- Performance Testing + QA Systems Certification + Products Certification

These were selected because CNAEL members agreed that each can stand alone as a viable alternative. Members noted that training should be a part of any program, and should be considered as an essential element. Members also noted that many of the alternatives were components of others and that several could be combined to create separate and viable alternatives. Federal pre-emption was eliminated as an alternative because it was not a practical solution in the light of states' rights. National guidelines were considered in the context of an administrative option, rather than as both an administrative option and a program alternative. For the purposes of characterizing the technical aspects of the alternatives, CNAEL agreed that the alternatives as listed above are in order of decreasing rigor.

CNAEL also identified seven different options for administering a program. These ranged from a completely centralized federal program to one owned and operated by the private sector with no federal or state participation. Presentations on current systems of operation were provided by the American Association of Laboratory Accreditation (a private accrediting body), the California state program, and the National Institute of Standards and Technology programs for Weights and Measures. The list was narrowed to three options:

- Oversight by the federal government with non-federal (state or private) implementation;
- Federal guidelines developed for voluntary adoption by state or private programs; and
- Administration by a private sector organization with federal and state government cooperation.

In making its selections, CNAEL concluded that centralized federal operation would be infeasible (largely due to resource constraints); that achieving state reciprocity without federal intervention is difficult but achievable (currently New Jersey, New York and California are working cooperatively); and that, a program operated entirely by the private sector is not likely to be accepted by the majority of states and, therefore, would do little to change the status quo. The status quo was deemed to be unsuitable due to its inherent problems. CNAEL agreed to the following refinements to two of the three administrative options:

- The option for federal guidelines should be assumed to involve some Federal oversight for implementation of the guidelines, and
- The options for federal government oversight with non-federal implementation should be assumed to include the possibility of delegating the oversight authority to a private sector accrediting organization. State represen-

tatives on CNAEL expressed concern about oversight of states' programs by a non-federal organization given the regulatory use of the data.
- The option for federal government oversight with non-federal implementation should be assumed to include the possibility of the states delegating the accrediting function to a private sector accrediting organization.

13.2.5
Elements of a National Environmental Laboratory Accreditation Program

CNAEL also defined the elements of a national environmental laboratory accreditation program. CNAEL accepted, with minor modifications, the elements of a program outlined by the International Standards Organization (ISO) Guides 25, 43, 54 and 55 on accreditation.

13.2.6
Scope of the Program

CNAEL also defined the scope of the program in terms of the types of laboratories, the methods/tests, and the environmental programs. The subcommittee used Federal environmental statutes as a starting point for defining the purpose of accreditation. Accommodations for technology advances or changes in measurement technology were considered to be pivotal in establishing a system which would be flexible enough to serve all parties. Agreement was reached that a national program should be applicable to all laboratories performing tests related to Federal environmental statutes, including public, private, and academic laboratories.

13.2.7
CNAEL's Conclusion and Recommendations

CNAEL's final report [13] was published in September 1992 and recommended the establishment of a national environmental laboratory accreditation program that has federal oversight and is implemented by the states and/or third parties.

13.2.8
Next Steps

The EPA Ad Hoc Panel on Laboratory Accreditation analyzed and evaluated CNAEL's recommendations and agreed with them. The joint recommendations of CNAEL and the Ad Hoc Panel were presented to the EMMC and EPA's Deputy Administrator. Since EPA was convinced of the advisability and feasibility of a national environmental laboratory accreditation program, the Ad Hoc Panel was instructed to develop a draft document describing a national environmental laboratory accreditation program. Because this can only be successful with the full cooperation of the States, EPA established a group comprised of state and EPA officials known as the State/EPA Focus Group to develop the details of a national program.

13.3
Program Developement

13.3.1
Setting Standards

To achieve the goal outlined by CNAEL to establish a national environmental laboratory accreditation program, uniform standards must be developed and adopted by all accrediting authorities. The State/EPA Focus Group has designed a system (modeled on a program operated by the National Institutes of Standards and Technology for weights and measures) which would provide an open forum for discussion and information exchange, while maintaining the responsibility for decision making to the regulatory bodies. An annual National Environmental Laboratory Accreditation Conference (NELAC), would be held which would include deliberations on the development of the standards and a voting session. In between the annual conferences, meetings would be held in various locations throughout the country enabling greater access at a reduced cost to the participants.

Proposed standards for environmental laboratory accreditation would be published in the Federal Register before both the Conference and the interim meetings to give the participants enough time to adequately prepare for the discussions. The initial set of standards which are proposed by the State/EPA focus group are based on ISO Guides 25 and 58, which were selected by CNAEL to promote international consistency. The standards include specific criteria on:

- quality systems,
- performance evaluation samples,
- on-site audits,
- legal authority and compliance monitoring,
- public relations and information dissemination,
- the accreditation process for the laboratory, and
- the policy and program structure for the Conference itself.

Under preparation are specific standards for the accrediting authorities.

13.3.2
Scope of the Program

Each program within EPA would either require or strongly recommend that only laboratories accredited as being in compliance with NELAC standards, be used for any tests conducted to serve EPA monitoring, enforcement, or other functions mandated by satutes and pursuant regulation. Some of the statutes are the Safe Drinking Water Act (SDWA), the Clean Water Act (CWA), the Clean Air Act (CAA), the Resource Conservation and Recovery Act (RCRA), the Comprehensive Environmental Response, Compensation and Liability Act (CERCLA), the Federal Insecticide, Fungicide and Rodenticide Act (FIFRA), and the Toxic Substances Control Act (TSCA).

Accreditation would be granted on a basis which is termed "field of testing". This is a four tiered system ranging from general laboratory requirements to specific method or parameter specifications. The first level of requirements would be general, encompassing such criteria as an adequate quality system, documentation of personnel training and qualifications, and directions for data storage and manipulation. The next level of accreditation would be the field of chemistry, biology, or field testing. The third level of requirements would be the traditional categories within those fields. For example, the chemistry portion would be further divided into inorganic, organic and other branches. The final level, reflecting the various requirements outlined in the separate EPA statutes (as mentioned above), would be a specific parameter, group of parameters, or method. For example, a laboratory might be accredited to perform Inductively Coupled Plasma methods on drinking water samples only. This laboratory, however, would be required to comply with all accreditation criteria covered under the general, chemical, and inorganic categories in addition to those specific to ICP analyses under the SDWA. The fourth level of requirements is most closely tied to the current regulations and would, therefore, be subject to significant changes based on regulatory revisions.

13.3.3
Federal Role and Responsibility

In accordance with the CNAEL recommendations the program, once developed, would be implemented by non-federal agencies with federal oversight. The Focus Group has recommended that the EPA perform the federal oversight function. This role includes:

- reviewing the accrediting authority's protocols, records, and performance,
- issuing a letter attesting to compliance with all relevant NELAC standards,
- evaluating and approving all state and federal laboratories,
- training of accreditors,
- management of the performance evaluation (PE) sample program, and
- management of a central data base.

The PE program would have a centralized unit in charge of production, distribution, and evaluation. Flexibility would be incorporated into the system to accommodate state run programs. The central data base would incorporate all state accreditation listings and would include data on PE samples, audit reports, and fields of testing (see section of scope for mor details on fields of testing).

In order to effectively operate such a program EPA would establish a National Environmental Laboratory Accreditation Program (NELAP) office. NELAP would represent the interests of all EPA national program offices, regardless of the statute covering their jurisdiction. To facilitate cooperation and communication between EPA and the Conference, two members of the NELAP staff would be ex-officio members of the Conference's Board of Directors. NELAP would also be charged with organization and coordination of the Conference.

13.3.4
Accrediting Authority Review Board

The EPA would establish an independent review board, the Accrediting Authority Review Board, composed of state and federal members. Thus purpose of this board will be to:

- review the process and procedures employed by EPA to evaluate the accrediting authorities' programs
- review the process and procedures employed by EPA to evaluate the federal and state laboratories,
- respond to complaints from accrediting authorities regarding EPA's consistency and conformance with the NELAC standards, and
- report findings to EPA and NELAC

The reports will aid NELAC to identify areas which are vulnerable to interpretation or which may induce conflict during application. The Board would not, on the other hand, address any complaints arising between a laboratory and the accrediting authority. These complaints may surface during the EPA's evaluation of an accrediting authority's program and therefore be indirectly addressed during the Board's review of EPA's procedures. However, the Board would only evaluate EPA's oversight function, not the accrediting authority's decisions regarding a dispute with a laboratory.

13.3.5
State Implementation

Laboratories would be accredited by state programs. In order to accommodate the regulatory responsibilities of the states, the decision was made that only states (or federal agencies with accreditation responsibilities) would have the authority to accredit laboratories. These accrediting authorities may select to use a third party to audit the laboratories, but would retain the decision making authority on whether accreditation should be granted.

The requirements and structure to obtain recognition as an accrediting authority under NELAC standards has not been selected. Not all states may be able, or want, to operate a comprehensive program due to legislative restrictions and administrative overhead. Many options exist and some of the proposed alternatives are described below (for the sake of simplicity the term accrediting authority and state are used interchangeably):

- the state would offer accreditation in all fields of testing using state personnel and/or a contracted third party to conduct the on-site audit,

- the state would offer accreditation in a limited field of testing, e.g. drinking water, and the laboratory could seek accreditation in complementary fields of testing from a second state, or
- the state would offer accreditation in a limited field of testing, but if a laboratory requests accreditation in fields of testing outside the capability of the state's program, a second state may accredit the laboratory for all fields of testing including those offered by the state of residence.

The first option which requires a comprehensive program provides the obvious advantage of a single source for all desired fields of testing, resulting in only one on-site audit. The drawback of this option is that some states, due to legislative, administrative, or technical reasons, are not capable of offering a program covering all fields of testing. This may be especially significant at the beginning of the national environmental laboratory accreditation program.

The second option, in essence, requires a laboratory to be accredited in the state of residence (primary state) for any desired fields of testing which are offered by the state. A second supplementary accreditation from another state source (secondary state) would be required for those fields of testing not offered by the primary state. The second on-site audit may be somewhat abbreviated because the initial audit by the primary state would have already evaluated the general criteria. Provided that the secondary state offers accreditation in all the additional fields of testing, a maximum of two on-site audits would be necessary.

The third option would permit the laboratory to seek accreditation in the state which best met their needs and could result in a single accreditation. The disadvantage of this option is that the primary state may experience difficulty in maintaining the viability of their own program if there are a large number of laboratories which select to have an accreditation performed by another state, threfore bypassing the primary state.

Other options can be envisioned and would be possible candidates for selection, including combinations of those outlined above. Further deliberations will ensue before a recommendation is made by the State/EPA Focus Group.

Each state is allowed to set fees which meet their own fiscal needs and responsibilities. A state could elect to charge a processing or licensing fee as part of the recognition process under reciprocity.

13.3.6
Reciprocity

As part of the demonstration of compliance with NELAC standards, any accrediting authority would be required to grant recognition of all laboratories accredited by a similarly qualified accrediting authority. Any accrediting authority not granting reciprocity would be judged to be out of compliance with NELAC standards. In this manner the NELAC standards would remain voluntary, and at the same time, would be the basis for a nationwide acceptance based on uniform accreditation standards.

13.4
Conclusion

The Unitd States Environmental Protection Agency (EPA) is charged with protecting public health and the environmental in the United States. In order to do this, environmental data must be gathered and analyzed. These data include information generated by many federal, state, and commercial laboratories from analyses of environmental samples. The quality of the analytical data must be known and documented to assure proper protection of public health and the environmental, as well as to ensure that costly pollution control and clean-up decisions are appropriate. A national environmental laboratory accreditation program is one step to assuring the quality of the data.

In 1994, EPA conducted a conference achieved consensus on a set of draft standards for a national laboratory accreditation program. These standards are to be reviewed for a final endorsement by the U.S. EPA Environmental Monitoring Management Committee in late 1995. If they are then endorsed, they will serve as the basis for a national laboratory accreditation program.

13.5
References

1. Stevenson R (1993) "Laboratory Accreditation: Is it Worth the Price?", American Environmental Laboratory, pp 6–8
2. Maxwell S (1993) "Industrial Update", Environmental Testing and Analysis, pp 28–33
3. American Council of Independent Laboratories (1985) "Accreditation of Environmental Testing Laboratories. An Independent Laboratory Perspective", pp 1–25
4. Harris P (1992) "Accreditation, CLP Heads Agenda for Analytical Laboratories", Environment Today, Vol 3, No 9, pp 3–28
5. Unger P (1991) "Environmental Laboratory Accreditation", American Environmental Laboratory, Vol 3, No 5, pp 24–25
6. Hyer CW (1991) "Directory of State and Local Governmental Laboratory Accreditation Designation Programs", NIST Publication # 815, pp 1–81
7. Berger H (1986) "The National Voluntary Laboratory Accreditation Program", American Laboratory, Vol 18, pp 8–10
8. Locke J (1987) "Laboratory Accreditation", Environmental Sciences and Technology, Vol 21, No 4, pp 332–333
9. Johnson G, Wolfe DW, Blacker S (1994) "QA Requirements for Environmental Programs", Environmental Testing and Analyses, pp 49–93
10. Hess E (1986) "Laboratory Accreditation – an Independent Laboratory Prospective", ASTM Standards News, pp 31–34
11. Stanger D (1989) "Evaluation of Accreditation Systems and Their Demands on the Independent Testing Laboratory", ASTM 1057, pp 78–84
12. Nadkarni RA (1993) "ISO 9000: Clearing the Hurdles on the Way to Quality Management", Today's Chemist at Work, pp 13–38
13. Mourrain JH et al (1992) "Final Report of the Committee on National Accreditation of Environmental Laboratories", United States Environmental Protection Agency
14. Edwards K, Trovato ER, Hankins J (1992) "Environmental Laboratory Accreditation: U.S. EPA's Efforts to Resolve the Debate", American Environmental Laboratory, pp 8–10

Appendix

Selection of some organizations working in the fields of accreditation, certification and measurement and testing within Europe (after B. Steffen, M. Wloka, S. Stobbe, BAM – Bundesanstalt für Materialforschung und -prüfung, Berlin, Germany)

acronym	name of organization address	year of foundation	task
CCQM	Comité Consultatif pour la Quantité de Matière R. Kaarls; Van Swinden Laboratorium bv; Postbus 654, NL-2600 AR Delft	1994	CCQM deals with – matters relating to the accuracy of quantitative chemical measurements and traceability to the SI, – establishing international traceability at the highest metrological level, – uncertainty statements in chemical measurements.
CITAC	Co-operation on International Traceability in Analytical Chemistry Dr. B. King; Laboratory of the Government Chemist; Queens Road, Teddington; GB-Middlesex TW11 0LY; UK	1993	CITAC is a worldwide international initiative built on work of existing organizations; CITAC organizes workshops open to all scientists with an interest in developing the international chemical measurement system
EAC	European Accreditation of Certification Paul M. Hewlett; NACCB; 19 Buckingham; GB-London SW1P 6LB	1991	Harmonization of the accreditation of certification bodies throughout Europe with the aim of acieving mutual recognition of accreditation bodies
EAL	European Co-operation for Accreditation of Laboratories Dr. R. Kaarls PO Box 29152 NL-3001 Rotterdam	1994	Fusion of WELAC and WECC (refer to them for task description)

acronym	name of organization address	year of foundation	task
EOTC	European Organization for Testing and Certification H. M. Jorg; EOTC, Rue de Stassart, 33; B-1050 Bruxelles; B	1990	Creation of a uniform European system for testing and certification
EQS	European Committee for Quality System Assessment and Certification Dr. Ing. K. Petrick; DQS, Burggrafenstraße 6; D-10787 Berlin	1989	Coordination and harmonization in the field of assessment and certification of quality assurance systems by third-parties
EURACHEM	Co-operation for Analytical Chemistry in Europe Dr. Maire Walsh State Laboratory Abbotstown, Castleknock IR-Dublin 15	1990	Promotion of the co-operation of analytical laboratories throughout Europe within the field of accreditation and quality assurance
EUROLAB	Organization for Testing in Europe Prof. Dr. C. Bankvall; Swedish National Testing and Research Institute; P. O. Box 857; S-50115 Borås	1990	Promotion of the co-operation of testing laboratories; Harmonization and confidence inspiration within European metrology; Promotion of mutual accept-ance of test results and quality assurance in the testing laboratory
EUROMET	European Collaboration in Metrology Kim Carneiro Danish Institute of Fundamental Metrology Anker Engelunds vej 1 DK-2800 Lyngby	1987	Promotion of co-operation between metrological institutions
ILAC	International Laboratory Accreditation Conference Dr. J. G. Leferink; NKO/ STERIN/STERLAB; Postbus 29152; NL-3001 GD Rotterdam	1977	Promotion of the exchange of experience between accreditation bodies and testing laboratories with the aim of an international recognition of test certificates and test marks and the facilitation of trade (realiz-ation of the GATT principles)
WECC	Western Europe Calibration Co-operation R. Kaarls; Van Swinden Laboratorium bv; Postbus 654; NL-2600 AR Delft	1975	Harmonization of the procedures of calibration institutes and frontier-crossing recognition of calibration certificates (fusion with WELAC into EAL, see there)

acronym	name of organization address	year of foundation	task
WELAC	Western Europe Laboratory Accreditation Co-operation	1989	Harmonization of accreditation of testing laboratories within Europe with the aim of mutual recognition (fusion with WECC into EAL, see there)
WELMEC	Western Europe Legal Metrology Co-operation S. Bennett; NWML – National Weights and Measures Laboratory; Teddington; GB-Middlesex TW11 ONH	1989	Harmonization and coordination of national and regional activities in all technical aspects of statutory metrology within Europe

Subject Index

Accreditation 15, 17, 91, 236
- bodies, requirements for 219
- policy 12
- procedure, general scheme 26
- program, US National Environmental Laboratory 255
-, decision on 224
Amount measurements 170
Assessment 222
Austria, the accreditation system of 28

Calibration 126, 204
- procedures, validation of 131
Cecking measuring 83
CCQM 261
CE labelling 9
CE mark 15
Certification 17, 91
- policy 12
Certified reference materials 196
- - -, use of 200
Certified values 199
CITAC 261
CITAC Document 25
Clinical analysis 207
Comparability 172, 176
Confidence interval 111
Conformity assessment 2
Control analyses 85
- chart 84, 87, 122
CRM 196

DANAK 35
Danish accreditation system 35
Decision, limit of 130
Detection, limit of 130
Distribution, normal 114

EAC 31, 261
EAL 31, 261
EN 45000 series 22
EN 45001 24
EN 45001 - EN 45003 77

Environmental analysis 206
- laboratories, accreditation of 247
EOTC 10, 262
EQS 262
Error, integration 100
-, materialization 101
-, random 108
-, systematic 106
Errors, types of 105
EURACHEM 24, 241, 262
EUROLAB 262
EUROMET 262

FINAS 38
Finnish accreditation service 38
Food analysis 206

Global concept 1
GLP 23, 229
GLP Certificate 235
Good Laboratory Practice 23, 229
Guidance 23

Homogeneity of reference materials 198

ILAC 262
Interlaboratory studies 88, 209
Irish Accreditation System 46
Irish National Accreditation Board 45
ISO 9000 series 20
ISO Guide 25 24
ISO/IEC Guide 25 22
Italian Accreditation System 40
Italian quality system 41
Laboratory performance studies 210
Limit of decision 130
Limit of detection 130

Machine capability 89
Means 116
Measurement uncertainty 99

NAMAS 65
National accreditation systems 28
NATLAS 65
Netherlands Accreditation System 51
New conception 2
NKO 51
Norway, national accreditation body of
    47
Norwegian accreditation 47
Notified body 7
NSAI 45

OECD 23
OFMET 30
Organizations 261
Outliers 115

PCBC 54
Personnel, qualifications 81
Polish accreditation system 54
Preparation of reference materials 197
Process capability 89

QA measures 83
QAS 23
QMS 23
Quality assurance 69, 77, 239
– assurance program 232
– assurance system 23
– assurance, statistical 118
– audits, internal 89
– control 70, 75
– control charts 122
– costs 74
– inspection 75
– management 70, 71
Quality Management System 23, 70
Quality Manual, compiling 78
Quality planning 70, 75
– policy 70, 71

Reference materials 87, 177, 179, 195
– –, certified 196
– –, definitions 196
– –, homogeneity 198
– –, preparation 197
– –, stability 198
– –, traceability of 183
– –, use of 200
Reference values 199
RM 196
Ruggedness 152
Russian system for analytical laboratory
    accreditation 57

Sampling 95
– procedures 98
SAPUZ 32
SAS 30
Sectorial committees 221
Sequential analysis 120
Sequential test plan 122
Significance tests 113
SINAL 41
SINCERT 41
SOP 233
Stability of reference materials 198
Standard deviation 116
Standards 19
Standards, European 77
Statistical quality assurance 118
Statistics 105
STERIN 51
STERLAB 51
Standard Operation Procedure 233
Surveillance 225
SWEDAC 60
Sweden, accreditation of laboratories in 60
Swiss accreditation service 30

Test control 83
– equipment 83
– instructions 85
– plan, sequential 122
Testing 17
–, attribute 120
Tests, significance 113
Total quality management 73
TQM 73
Traceability 159, 176
– of reference materials 183
–, criteria for 187
–, purposes 184
–, definitions 187
Trend 115

UKAS 63
Uncertainty ranges 111
Uncertainty, measurement 99
United Kingdom Accreditation System 65
USA, accreditation in 247

Validation 135, 179
– of calibration procedures 131
–, basic 136
–, definitions 138

WECC 262
WELAC 24, 263
WELMEC 263

Springer-Verlag
and the Environment

We at Springer-Verlag firmly believe that an international science publisher has a special obligation to the environment, and our corporate policies consistently reflect this conviction.

We also expect our business partners – paper mills, printers, packaging manufacturers, etc. – to commit themselves to using environmentally friendly materials and production processes.

The paper in this book is made from low- or no-chlorine pulp and is acid free, in conformance with international standards for paper permanency.